CANAL COUNTRY

Utica to Binghamton

Text
Emily Williams

Photography
Helen Cardamone

Foreword
David M. Ellis

P. V. Rogers Professor of American History
Emeritus, Hamilton College

Also by Emily Williams and Helen Cardamone
STAGECOACH COUNTRY
CHERRY VALLEY COUNTRY

Library of Congress Catalog Number 82-090374

Copyright 1982 by Emily Williams and Helen Cardamone

First Edition

Printed in The United States of America

All Rights Reserved

ISBN 0-9608330-1-3

For Joe, in loving memory,
and in gratitude for encouraging us
to start the series of books.

Foreword

David Maldwyn Ellis, P. V. Rogers Professor of American History Emeritus, Hamilton College.

If the Erie Canal made Utica the commercial emporium of the western district of New York, the Chenango Canal transformed it into a manufacturing center by the late 1840s. Near the juncture of the Chenango and Erie Canals woolen and cotton factories sprang up utilizing Pennsylvania coal freighted northward through a hundred or more locks from Binghamton to Utica.

The same team that brought us such delightful books as *Stagecoach Country* and *Cherry Valley Country* has provided another fine addition to the history, folklore, and architectural development of central New York. Helen Cardamone and Emily Williams have discovered a successful formula: select a region of manageable size and charm, capture its important public and private buildings in a hundred photographs, and interlard among the pictures sketches of significant personalities. This time their task was more difficult because of the scarcity of written sources. I can think of no foreign or American observer who has left us accounts of a journey from Utica to Binghamton by land or by water. Fortunately each one of us can ride along the route of the Chenango Canal and sample the flavor of such pretty villages as Clinton, Hamilton, Norwich, Oxford, and Greene. This book should tempt many readers to make such a sentimental journey through one of the loveliest landscapes in the state.

Emily Williams has chosen to present history through biography. In Utica she chose Alfred Munson whose career spanned the city's rapid development from a stage coach center to canal port to factory city. Munson founded a mill to make burrstones essential for the age of waterpower. He operated a foundry to make tools and later invested in textile mills and railroads. His fortune continues to enrich citizens through the family legacy of parks and the Munson Williams Proctor Institute. Louisa Barker of Clinton operated academies for girls and built a Gothic style house on Chestnut Street. Thomas Dean's story is a fascinating account of his Quaker concern for the Brothertown Indians who were demoralized by the evils and diseases of white men's civilization. Dean worked hard to ease their transplantation to the Oneida reservation near Green Bay, Wisconsin.

Esek Steere, another Yankee, created a prosperous business in Hamilton. Although childless, he and his wife took a keen interest in education and Steere helped capture for their village Madison University, now Colgate. How many people know of the Morse family of Eaton who used waterpower as the basis of their manufacturing empire: grist and sawmills, foundries, and even a plant to make portable steam engines well regarded by prairie farmers. When farm machinery plants followed the westward movement, the Morse empire withered. George Page of Earlville, another obscure individual, became a substantial merchant until the railroad undermined his business which was tied to the Chenango Canal.

I am sure that few have heard of Hosea Dimmick, locktender of Norwich who lived close to the Maydole Hammer Factory. Better known was "Honest" John Tracy, Oxford lawyer, who won election to the Assembly and became Lieutenant Governor in 1832. Another politician, Frederick Juliand of Greene, fought in the state Senate for the extension of the Chenango Canal south of Binghamton. Charles Samuel Hall of Binghamton, a Yale graduate and author, won honors at the bar and encouraged many civic improvements.

These capsule sketches tell us a good deal about the marketing of agricultural products, the impact of the canal upon communities, the rise and decline of sustaining industries geared to waterwheels, the immigration of Yankees and the Irish, and the development of manufacturing. The photographs of buildings dating to the middle third of the nineteenth century present a graphic image of gracious, almost idyllic times when ambitious entrepreneurs erected mansions and enjoyed a quality of life worthy of the beautiful landscape.

I congratulate the authors once more for providing in *Canal Country* a re-creation of a vanished age through the medium of camera and pen. One can only hope that they will look around for new worlds to conquer.

Contents

The Chenango Canal

The Chenango Canal, a silvery ribbon through some of the most beautiful countryside in the Empire State, tied central New York to the Southern Tier. Caressing the docks of 18 communities in the 97 miles between Utica and Binghamton, the canal joined them to the prosperous Erie Canal and held them in a companionable web of transportation which brought comfortable prosperity. The canal rippled through lush valleys and lapped against fields and pastures with the heartening message that products of the fertile farms could move to market with ease. The waterway slashed freight rates from $1.25 to 25¢ a ton for the Binghamton-Utica run, moved the freight in less than half the previous time, and pulled the Susquehanna area into workable communication with the center of the state. In a crescendo of activity during the fifties, the canal's coal barges fed riches to the mills of Utica.

The canal passed through communities where pillared Greek Revival houses gleamed white beneath the soaring spires of dusky stone churches. Pretty, little frame houses snuggled beside big Victorian mansions. Nearby, massive stone farmhouses presided over the gently rolling hills.

From Clinton, shortly after its junction with the Erie at Utica, the canal entered the lovely valley of the Oriskany Creek and ascended to the watershed level near Hamilton where a glacial moraine separated the tributaries of the Mohawk from those of the Susquehanna. At the summit elevation, seven lakes and ponds with their network of canal feeders assured a generous water supply. From the summit down to the Susquehanna, the canal followed the smooth valley of the Chenango River, known to the Indians as "Place of the Bull Thistle," or "Pleasant River which flows through a beautiful and fertile country."

Like the bull thistle, the roots of the Chenango Canal's history ran deep, bedded in the days when men dreamed of a vast inland waterway system which would unite the St. Lawrence River with Chesapeake Bay. The dream persisted during the years when enthusiasts strained for permission to build a canal from the Hudson River to Lake Erie. In 1814, shortly before the struggle reached its successful climax, a commissioner remarked in the Legislature at Albany that "It will not be difficult to extend that communication to the fertile vales watered by the Susquehanna and its wide spreading branches."[1] The Erie opened in 1825, and within the next few years New York State expanded its canal system and built nine lateral canals which nourished and drew strength from the magnificent main artery.

The Chenango Canal, a lateral of the Erie, opened in May, 1837. From Utica to the summit level, canal boats rose 706 feet through 76 locks in 23 miles. Boats descended 303 feet and eased through 40 locks in the 74 miles from the summit to the Susquehanna at Binghamton. The masonry locks, said to be the finest in the state, were 90 feet long, 15 feet wide, designed to hold a single canal boat about 14 feet

---

[1]Noble E. Whitford, *History of the Canal System of the State of New York*, Albany, 1906, p. 42.

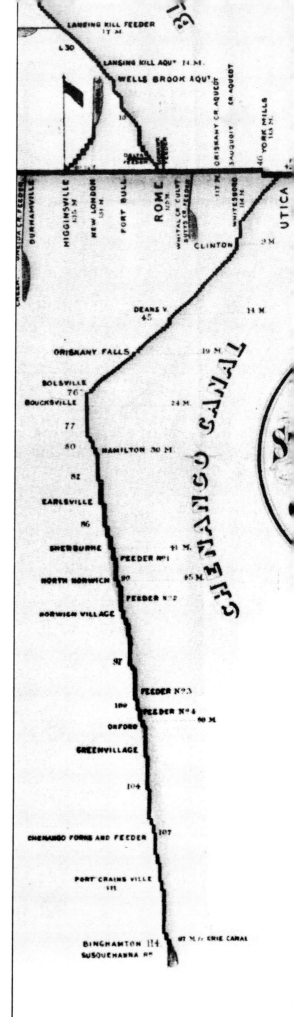

floating homes and barns as well as cargo carriers. The captain, cook and steersman ate and slept in the cabin at the rear. Cargo filled the hatches in the waist of the boat. Mules or horses munched their hay and grain in the stable at the front. At night the hoggie crept into a cubby-hole under the bow.

A captain usually worked each team for about six hours at a time. At the end of the shift he ordered hoggie and steersman to stop beside the towpath. He dragged out the gangplank or "horse bridge." The hoggie led out the fresh team, secured the tired animals in their stable, and the freighter resumed its slow, steady progress.

Most of the freighters were painted white each spring, and showed very little decoration except at the stern where, just above the rudder, the name and home port glittered in bright colors. On the Chenango, steersmen leaned against long wooden tillers above a rich variety of names and places . . . "Duke of Wellington, Albany" . . . "Oneida Bank, Utica" . . . "Lame Indian, Elmira" . . . "Thomas, Buffalo" . . . "Safety-Fund, Greene" . . . "Hamilton Snuff, Binghamton" . . . "Fair Play, Norwich" . . . "Brandywine, Oxford" . . . "North River, Sherburne" . . . and hundreds more.

A few packet boats carried passengers along the Chenango Canal. In the 1850s one boat line offered thrice-weekly service between Binghamton and Hamilton. For many years the "Lillie" and the "Norwich" and other packets provided daily service from Norwich to Binghamton and return, leaving Norwich at 6 a.m., arriving at Binghamton between 6 and 8 p.m.

On nice days passengers crowded the roof of the packet and sat on the straight chairs and benches placed there for their pleasure. Ladies in voluminous skirts and huge bonnets raised their bright parasols against the sun. Gentlemen in top hats enjoyed their ease, but when the packet approached one of the bridges which carried highways or farm roads across the canal, the captain's command of "Low Bridge!" shattered the journey's tranquillity. As on the Erie Canal, bridges hung close to the water. Often they squeezed the space between the top of the packet and the bottom of the bridge to a clearance of about two feet. At the captain's shout the crowd of passengers bent low, like a field of ripe grain flattened by a gust of wind.

The Chenango Canal carried only a tiny fraction of the traffic which surged along the Erie. While a lock tender on the Erie could record about 200 lockages in 24 hours at the height of the season, a lock tender on the Chenango seldom eased more than a dozen boats through his lock on a busy day. Nevertheless, a remarkable amount of freight moved on the Chenango. 4,000,000 feet of lumber moved north from Binghamton in the 1840 season. The Chenango Agricultural Society reported that 1,035,256 pounds of cheese were shipped from the Norwich docks in 1849. Canal freighters carried hay, corn, oats, potatos, hops, apples, eggs, pianos, hop poles, empty firkins, wagons, leather, tallow, limestone, iron ore, pig iron, whiskey, kettles, furniture, crockery, salt, plaster, steam engines, coal and many other vital, welcome products.

For 40 years the Chenango Canal served as friend and helper to those who lived near its towpath and berm, but as it grew older, ills and troubles overwhlemed the pleasant waterway. The Erie was enlarged, and New York State boats were built bigger and loaded heavier. At the same time, the Chenango locks weathered and shrank inward a few inches, as if cringing in apprehension. The new boats could hardly push through. They injured the old locks and forced disastrous remedies. Lock tenders put up planks like spillway boards to raise the level of the water. When the gates opened, the overflow gushed from the locks and crumbled the canal's fragile banks. Maintenance costs rose and traffic dwindled.

Deterioration was not the only threat. Remorselessly, during the 1860s, railroads pushed their tracks through the valleys of the canal route. By 1870 a railroad line linked Utica and Binghamton. Freight could move by rail from the Susquehanna to the Erie in three hours rather than three days, and at no greater cost. Like greedy vultures the black iron monsters grabbed all the canal's business. Mortally wounded, the Chenango Canal died of starvation, and was abandoned after the 1877 season.

Canal Country follows the route of the Chenango through the communities close to the old canal. The lovely photographs show buildings which existed in their charm and beauty when teams trod the towpath. The text identifies the original owners. Today's Routes 12 and 12B follow the canal route closely, lead motorists through the canal communities, and reveal many traces of the waterway which brought rich opportunities to a gracious swath of the Empire State.

wide and 75 feet long, loaded to a maximum of 65 tons.

Teams of horses or mules pulled scows, lakers, packets and freighters along the New York State canals, with the team hitched tandem, one animal ahead of the other, at the end of a tow rope about 200 feet long. The driver or "hoggie," who usually was a boy, rode the rear animal or trudged the towpath behind the team.

The towpath ran along one side of the canal only. When boats met or passed each other, the hoggies and steersmen carried out a maneuver as precise as a minuet. The hoggie of the downstream boat moved his team to the outside of the towpath and halted. His steersman guided the boat to the opposite side of the canal, called the "heelpath" or "berm." Their rope went limp and sank to the bottom of the canal. The oncoming team picked its way over the length of inert rope which lay like a snake in the towpath. Its boat hugged the towpath side and swept along without slackening.

The snub-nosed freighters, called "ballheads" or "bullheads," were

Alfred Munson built the Tuscan villa at 318 Genesee St. for his daughter and her husband. Construction started in 1850.

ALFRED MUNSON

On June 5, 1823, Alfred Munson with Elizabeth, his bride, arrived in Utica, where the bridegroom counted on making a fortune. The village was the most important transportation center west of the Hudson River port of Albany. Settlers climbed down from wagons, stepped out of stages, and some disembarked from Mohawk River boats. They lodged and bought supplies before starting west on the Seneca Turnpike. Travelers came and went on nearly 20 stagecoach lines which radiated from the busy village. Already, boats glided along the fringe of the village on the first section of the new Erie Canal. Completion of the entire canal was expected in two years.

Alfred was 30 years old, and his demure, pretty bride was five years younger. Among the confident young men who crowded the streets, hurried to the stagecoach offices, strode to the mercantile houses and the banks, the thin six-footer stood out with distinction. He had a long face, dark hair, finely chiseled features, an aquiline nose which was much too long, and very bright, dark eyes.

Six days earlier the Munsons had been married in Connecticut. Like a gambler cashing in his chips, the bridegroom had turned his assets into money before leaving for New York State. He had sold his share of the family farm, grist mill and saw mill for a total of $2900. This gave him an ample stake for investment in an age when laboring men brought up families on wages of $1 a day.

Almost immediately, Alfred Munson made his first investment in Utica. He rented a basement at the corner of Liberty and Hotel Streets, and he began to manufacture high-quality buhr millstones. He succeeded brilliantly, and soon had to move to larger quarters. For over 150 years the business has moved and changed with the times. The Munson Machinery Company of Utica is a proud descendant today.

In the years after his arrival the shrewd adventurer found rich opportunities to match his high aspirations. He invested in steamboats on the Great Lakes, and bought packet boats on the Erie Canal. He put money in the new railroads and served on their far-seeing boards of directors. As president of the Oneida Bank he guided the Utica institution through the perils of internal corruption and external panic. He speculated in real estate in Baltimore, Maryland and in Michigan. He controlled iron works at Baltimore and at Franklin Springs near Clinton. He purchased leases for huge, rich coal fields in Pennsylvania.

While Alfred Munson's iron will pushed his businesses to prosperity, he struggled against his own fading health. Tuberculosis sapped his strength, and the cold, damp, Utica winters tormented him with acute suffering. For many years he went south in the winter months and sought vigor and stamina in the sunshine of Florida and Cuba.

During his long absences from Utica he worried about his wife Elizabeth and their children, Helen and Samuel Alfred. "I hope," he wrote Elizabeth one year, "the children will behave well and obey you. Alfred must not associate with the wild boys, must keep at home with William [the coachman] and William can carry you to ride at any time . . . and also the children."[1]

The year 1845 brought a time of unforgettable change in Utica. The village had expanded, had become a city in 1832 and had reaped riches from its location on the Erie Canal. Yet the canal had hurt the city as well, for it robbed Utica of its preeminence as a hub of transportation. Boats filled with freight and passengers glided past, without stopping. Taverns, stagecoach lines, mercantile houses, banks felt the dwindling of their old, rich trade. While cities in the western part of the state were growing fast, Utica's prosperity was slipping.

In July, 1845, the *Utica Gazette* ran a series of articles which pointed the way to recovery. The articles admonished Utica to abandon its dependence on banking, transportation and mercantile business, and turn to manufacturing. Of course, the articles conceded, there was hardly any water power available in the flat plain and former marsh where the city stood, but times had begun to change. Factories no longer needed rushing streams and dripping water wheels. Coal could power machines, and the fuel could reach Utica via the Chenango Canal.

The businessmen of Utica sprang to action. They sent a committee to Rhode Island to study the steam mills which had opened there just recently. The committee members were impressed, and they brought back an extremely enthusiastic report. While water power was notoriously vulnerable to changes in the water supply of streams, they noted, neither drought nor ice could interrupt the flow of coal from the big black piles to the boilers. Steam mills could be built in cities, very advantageously. Products could be shipped to market on nearby waterways, and, furthermore, labor was cheaper and more plentiful in the cities than in the country. Utica was a prime site for steam mills, because the Chenango Canal, like a life-line, could bring them the needed tons of coal.

Within the next two years the businessmen of Utica organized three textile mills, to be powered by steam. The chance to build revolutionary new steam mills seemed to delight Alfred Munson. Although his health was ebbing steadily, he grasped the lead among the Uticans who rushed to invest in the new mills and to serve on their distinguished boards of directors.

Alfred Munson headed the Utica Globe Mills, built in 1847. He was president of the Utica Steam Cotton

---

[1] Alfred Munson, MS, Oneida Historical Society, COR. 6.

Mills Company which, in 1849, built its first mill on Court Street at Chenango Canal Lock #1. "Buildings, engine and machinery [were] all of the best kind, neither money nor time having been spared in their erection and manufacture . . . containing the latest improvements . . . an ornament to the city."[2]

Soon after the opening of the steam mills, Alfred Munson's daughter Helen married the scholarly and attractive lawyer, James Watson Williams. In 1850 Mr. Munson started construction of a house for the young couple. This is the lovely Fountain Elms which stands today on Genesee Street.

Alfred Munson became enormously wealthy. He was thought to be the richest man in the county, but he did not hoard his wealth nor begrudge his time to worthy causes. He served on the boards of directors of the city's leading charitable and educational organizations, and he made generous financial contributions to them and to Grace Church.

Four years after Helen's house was begun, her indomitable father faced approaching death. "I am not as well as I was the first of last week," he wrote to his physician. "I am fearful that you have not fully realized this point in my disease. It is the main one and that keeps me down, and unless we can find relief must terminate my life.

"I am thus plain that you may fully understand the work we have to do to save me."[3]

Their work could not save him. Alfred Munson died in May, 1855. At his death Utica was starting on a new period of prosperity launched by the steam mills and fueled with coal brought by the Chenango Canal. In the next decades the canal was destined to be driven out of business by the railroads, but the steam mills were destined to rise to prodigious importance. Ultimately, they made Utica the knit-goods center of the world.

Alfred Munson's large fortune enriches the city of Utica still, through the famed ornament to the community, the Munson Williams Proctor Institute.

---

[2]Pomroy Jones, *Annals & Recollections of Oneida County,* Rome, 1851, p. 615.

[3]Alfred Munson MS file, Oneida Historical Society, COR. 8.

The brick mansion at 711 Herkimer Rd. was built about 1815 by Gen. John G. Weaver, a commander of militia in the War of 1812. The bricks came from the large Weaver brickyard on the north bank of the Mohawk River.

Gen. Weaver was descended from George J. Weaver, one of the Palatinate Dutch pioneers who founded Deerfield before the Revolutionary War. Gen. Weaver and his wife Charity are buried in the First Baptist Church cemetery, Herkimer Rd.

The house at 1143 Herkimer Rd. was built about 1840. It is a fine example of the Greek Revival style, and is noted for its beautiful temple portico.

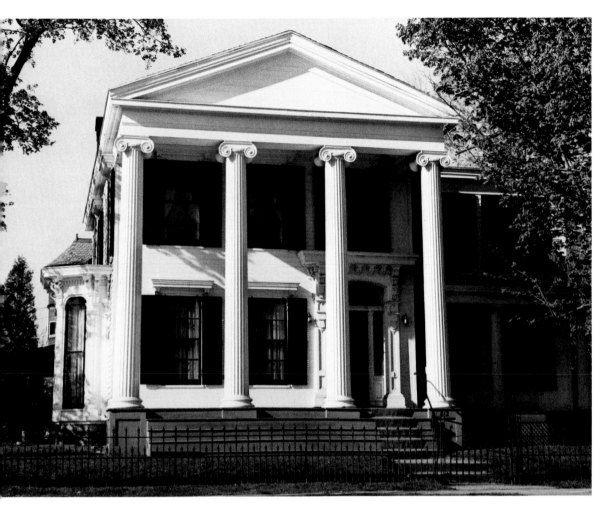

Morris G. Dickinson, wholesale and retail druggist, built the house at 212 Rutger St. in 1845.

Utica Psychiatric Center, Whitesboro St., was built by the State of New York in 1843. This was the first institution in the state to provide humane treatment for the insane. The staff rejected the old methods of brutal mistreatment and pioneered ways of trying to help the insane to recover. Dr. Amariah Bingham, the first Superintendent, founded the American Journal of Insanity, forerunner of the modern American Psychiatric Journal.

The building with its portico of Doric columns is said to be one of the world's finest examples of Greek Revival style.

The Italian villa at 1 Rutger Park was designed by the famous architect Alexander Jackson Davis and built for the banker John Munn in 1854. After starting his banking career in Utica, Mr. Munn moved to Mississippi where he made his fortune. He married a southern lady and brought his family back to Utica during the 1840s.

St. John's Church, corner of Bleecker and John Streets, organized in 1819, was the first Roman Catholic Church built west of Albany and the fourth built in New York State.

Two beloved Uticans, John C. Devereux and his brother Nicholas, are inextricably interwoven with the history of St. John's. Prominent merchants, bankers and industrialists, they had come originally from Ireland where their family estates were lost in the Irish Rebellion of 1798. Before the church was built, priests came from Albany several times to celebrate Mass in the home of John C. Devereux at the corner of Broad and Second Street. When the church was organized, the Devereux brothers served on the first board of trustees, and they were the largest contributors to the building fund.

In 1834 the brothers brought three Sisters of Charity to Utica and opened St. John's Orphan Asylum and School for Girls, in a small building next to the church. The school was the predecessor of Utica Catholic Academy.

The congregation grew too large for the original church, and in 1836 a new church was built, across Bleecker Street from the first building. This church was demolished and the present church was built on the site in 1872. The twin spires were added in the 1890s.

In 1829 or 1830, Philip Hooker of Albany designed the house at 3 Rutger Park for Judge Morris S. Miller, husband of the former Maria Bleecker of Albany. Judge Miller, prominent lawyer and Congressman, managed the large, Utica portion of the Bleecker estate for his father-in-law, Rutger Bleecker, one of the original patentees of Utica. "Millers' Seat," as the property was called sometimes, stood on land inherited by Mrs. Miller from her father. The property extended originally from Howard Ave. to Dudley Ave. and from Rutger St. to South St.

Judge Miller died soon after the foundations were laid. The house was completed according to his plans by Mrs. Miller and her son Rutger Bleecker Miller, Utica lawyer and Congressman.

For decades this house was associated with illustrious guests. The Millers entertained President Martin Van Buren, Gen. and Mrs. Winfield Scott, Gen. Lewis Cass, Sen. John C. Calhoun, Mrs. Alexander Hamilton. In 1843 a wedding reception was held here for the lawyer, later U.S. Senator, Francis Kernan and his bride Miss Hannah Devereux, daughter of Nicholas Devereux of Utica.

Among subsequent owners of the house were the Thomas R. Walkers, and their guests included President Millard Fillmore and Samuel F. B. Morse who was a cousin of Mrs. Walker's.

Sen. and Mrs. Roscoe Conkling bought the house in 1868. The most brilliant gathering during their tenure took place at the reunion of the Army of the Cumberland held in Utica in September, 1875. The Conklings entertained President and Mrs. Ulysses S. Grant at this time.

The house was purchased by Nicholas E. Kernan in 1894.

For many decades the house at 286 Genesee St. was the home of the Kellogg family. The Spencer Kelloggs lived here from 1842. Mr. Kellogg, a dealer in foreign and domestic dry goods, was mayor of Utica in 1841, director of the Bank of Utica when it was organized, member of Utica's first School Board. In 1845 Mr. Kellogg served on the committee which visited the new steam mills in New England. This committee's momentous recommendations spurred the building of Utica's first steam-powered textile mills. These were fueled by coal transported on the Chenango Canal.

Orsamus Matteson, one of Utica's most prominent lawyers, built the house at 294 Genesee St. in 1866. As a young man he had studied law in the office of Greene C. Bronson and Samuel Beardsley. Later he became a partner in the firm of Beardsley & Matteson. He served as City Attorney, 1834 and 1836. He was a member of the U.S. Congress 1849–51 and 1853–59.

After Mr. Matteson's death in 1887, William M. White bought the house and members of the White family lived here for two generations. Since 1919 the property has been owned by the Catholic Women's Club.

In the early 1840s the Jared Eliot Warners lived in the house at 296 Genesee St. Since 1817 Mr. Warner had been proprietor of the drug store on the northwest corner of Bagg's Square, the first drug store in Utica. Upon his retirement in 1867 the property was purchased and remodeled extensively by Mr. G. Clarence Churchill.

Utica's noted businessman, Charles Millar, bought the house at 1423 Genesee St. in the 1860s, soon after it was built. Mr. Millar and his wife the former Jane Quait had emigrated from England in 1835 and had settled in Utica in 1838. Trained as a master builder, Mr. Millar became a very successful building contractor in Utica. In 1861 he purchased the wholesale tin, plumbing and steamfitting business which still bears his name. His son Henry W. Millar joined the business in 1866, at which time the firm became Charles Millar & Son. The Millars celebrated their golden wedding aniversary in this house in 1883.

A tavern built probably in the 1790s forms the earliest portion of the stately Greek Revival house at 2108 Genesee St. In the 1830s the present residence was built for Daniel Mason, builder of the High Bridge over the East River at New York City. The house has a pedimented roof and a two-story portico supported by Ionic columns. The wing and portico on the south side were added about 1900.

The house stands on land which Daniel Mason's father, "Squire" Arnold Mason purchased about 1795 on the Genesee Rd., principal route from Utica to the West.

New Hartford

The house at 14 Oxford Rd., built probably in the 1840s or 1850s, stands on land owned originally by Gen. George Washington and Gov. George Clinton, and purchased in 1790 by Jedediah Sanger, New Hartford's pioneer industrialist and land-holder. Mr. Sanger's purchase covered a large part of today's village. He subdivided the tract and during the early 1800s he sold building lots on Oxford Rd., called "Chenango Turnpike Road" in some of the early deeds. Both names for the road revive the memory of a long-forgotten company, formed in 1801, which proposed to build a turnpike from today's New Hartford to the township of Oxford, Chenango County.

Walter S. Eames, grandson of Jedediah Sanger, built the house at 76 Oxford Rd. about 1830. He was one of nine children born to John Eames and his wife the former Sally Sanger. For many years during Walter Eames' childhood, the family lived in a small cottage which later became a wing of the big house.

The Eames gave a gala housewarming party when the big house was finished. A band played in the railed enclosure on the roof while guests promenaded the grounds.

Samuel Hicks, original owner of the house at 18 Oxford Rd., settled in New Hartford in 1804 and served as manager of the Eagle Mills near Clayville. The house was completed in 1826, and it is a splendid example of pure Georgian architecture. The doorway is especially noteworthy. Although many doorways of the period were too wide for their height, this doorway has proportions which have been called perfect.

In 1791 thirteen New Hartford pioneers gathered in Jedediah Sanger's barn and organized a congregational church. Mr. Sanger donated a large plot of land, and the church was built between 1793 and 1796. It was the first church building in New York State west of Herkimer. Rev. Dan Bradley, the first pastor, received a salary of £ 100 per year, payable one-third in cash and two thirds in produce at current prices, as well as 30 cords of good firewood per year. The church became Presbyterian in 1802.

Remodeling was carried out in 1827, 1851, 1871, but the original frame structure was retained.

The house at 72 Paris Rd. is one of the oldest dwellings in the village. The original owner was probably Ashbel Andrews who had purchased the land in 1794. The house was built in the very early 1800s.

The original farmhouse at 116 Oxford Rd. is believed to have been built in 1792 and was bought the following year by Eli Butler, Sr. when he moved his family from Connecticut to the Sauquoit Valley. This became the homestead of the Butlers, occupied by Eli, Jr. and his family and then by Eli's son Morgan and his wife the former Marianne Howard of Frankfort.

Before 1857 Morgan Butler transformed the house into a model of the Gothic Revival style. The massive central chimney was removed and two new ones were built to serve the new, shallow fireplaces. The house was lengthened at either end by one-and-one-half story additions, each with an oriel window and French doors downstairs and circular windows upstairs. A huge gable with Gothic lancet win-

dow was built in front of the attic. Porches were added in the front and rear, and all the eaves were decorated with scroll-sawn bargeboards.

In 1796 Nathaniel Seymour Andrews bought a parcel of land from Ashbel Andrews and in 1798 he built the house which stands today at 113 Genesee Street. Probably the oldest residence in New Hartford, it is a splendid example of post-Revolutionary architecture. In 1811 John H. Lothrop, Yale graduate, farmer, lawyer, editor, merchant, banker, Hamilton College trustee, bought the property. He was a man of social tastes and his wife Jerusha, a daughter of Rev. Samuel Kirkland, was known for her "wit, beauty, vivacity and vigor."[1] The Lothrops entertained the most prominent citizens of the vicinity at their home. From 1837 to 1937 the fine house was the property of the Palmer and French families.

[1]Bagg, Pioneers of Utica, p. 160.

Needham Maynard, one of the earliest settlers of today's Town of New Hartford, built the house on Middle Settlement Rd. about 1787. A close friend of the pioneer, Hugh White, Mr. Maynard is believed to have come from Connecticut.

The house is very noteworthy as a rare and extremely well-preserved example of Colonial architecture. It is built of timbers over two feet in diameter. The massive chimney, with flues for six fireplaces, is made of imported Holland bricks.

The house at 80 New Hartford St., New York Mills, known for many years as the "Wilcox Place," was built probably by Abel Wilcox in the early 19th century. The site is about a quarter of a mile from the route of the Chenango Canal.

The "Wilcox Place" stands on land which Abel Wilcox purchased from Brig. Gen. Oliver Collins in 1793. Gen. Collins, leader of the militia of Oneida, Madison, Herkimer, Jefferson and Lewis counties in the War of 1812, succeeded Gen. Jacob Brown in the command at Sackets Harbor. Gen. Collins emboldened his officers with stirring orders such as "Every inch of American soil is held sacred, and . . . no invader will be permitted to set foot in it. . . ."

The home of Louisa Barker, Chestnut St., was built about 1850.

LOUISA BARKER

In the early 1800s when Louisa Barker was a little girl in today's Franklin Springs, she played in cleared fields which lay like a thin coverlet above the deposits of rich red hematite iron ore. Sometimes men scooped the red soil, threw it on heavy ore wagons, and big teams of oxen hauled slowly through the settlement of Franklin and plodded toward the blast furnaces several miles away. When Louisa became a young lady, the paths and rutty roads turned to mud when it rained, and sometimes her long skirts were stained red on the hems.

While Louisa grew into an attractive young woman, nearby Clinton changed from a little settlement of mill-owners and artisans into one of the most fascinating and unusual educational centers in New York State. Since 1812 Hamilton College, an outgrowth of the Hamilton-Oneida Academy founded by Samuel Kirkland in 1793, had stood on the hill overlooking the village. Beginning about 1815, a variety of private secondary schools for both girls and boys opened in the village itself. At one time there were so many schools in Clinton that the place was called "Schooltown."

In 1840, when she was 32, Louisa became preceptress at one of Clinton's three boarding schools for girls, the Female Department of the Clinton Liberal Institute. The school belonged to the Universalists, believers in improvement and progress, individual freedom and the essential goodness of mankind. The Female Department was an expression of their beliefs, for it was a girls' school run in conjunction with a boys' school, a novel idea for the times. The Female Department occupied a small frame house at 14 Utica Street, about a block from the big stone building where the boys lived and studied. Louisa ran the school very successfully. Within 11 years the Female Department outgrew its little building, changed its name to White Seminary, and moved to a large, new, pillared edifice on Chestnut Street.

Big challenges confronted Louisa in 1850. She must guide the new seminary which would have an enrollment of 50 young ladies and would be the most elegant boarding school in Clinton. Furthermore, changes portended at the Franklin ore fields, and drew her into an exciting business venture.

Before 1850, the owners of the ore fields had carried their iron ore to villages such as Taberg, Westmoreland and Constantia where charcoal-fueled blast furnaces smelted the ore into pig iron. In the late 1840s, while scooping their ore, owners began to see canal boats gliding past the fields, carrying coal from Pennsylvania to the new steam mills at Utica. Fresh as sunrise, a new idea swept through Franklin. The local people decided to build a blast furnace. Using coal brought by the canal they would smelt their ore into pig iron, at their own village beside the Chenango Canal. They banded together in a stock company, the Franklin Iron Works Company. The 12 original stockholders invested a total of $16,000. Louisa was the only woman among the stockholders, and she served on the original board of directors.

The stockholders trusted in the success of their excellent product. Franklin ore was rich and pure. "The ores of this region," one historian wrote, "when used alone make the finest of castings for ornamental purposes . . . in their molten state they flow like water and fill up every part of the mould with perfect nicety . . ."[1] The molten iron flowed into hundreds of the glossy castings which decorated Victorian stoves. It helped make weights for scales, railroad rails, chairs, spokes, cannonballs.

Within a couple of years the original investors lost control of the iron works. A Utica group headed by Alfred Munson bought a major interest in the company, increased its capital and enlarged the blast furnace.

The Franklin Iron Works became one of the Chenango Canal's best customers. Boats hauled coal from Pennsylvania to fuel the giant blast furnace, and they carried away the pig iron which was smelted. Franklin ore became so well known that the demand outran the capacity at the local furnace. The collector's office at Hamilton recorded boats of ore moving south to Binghamton, "The Lame Indian, 30 tons ore" . . . "Hamilton Snuff, 50 tons ore" . . . "Dancing Feathers, 20 tons" and many, many more.

While the new group enlarged the Franklin Iron Works, Louisa enlarged her own role in the schools of "Schooltown." The White Seminary, she thought, seemed too small. She formed a partnership with Miss Chipman and the two women built a school of their own. This was the big, twin-towered, two-story Home Cottage Seminary on Chestnut Street. The new school opened in 1854 and reached an enrollment of over 90 girls. Soon Louisa grew dissatisfied again. She decided the school was too big. The Home Cottage Seminary was sold, and it became Houghton Seminary, perhaps the most famous of all the Clinton Schools. Promptly, Louisa opened the little Cottage Seminary on College Street. Here she limited her pupils to 15 boarders and 15 day students.

Home Cottage Seminary, Cottage Seminary and the Iron Works have vanished. The Female Department's original building still stands at 14 Utica Street. Pillars salvaged from the White Seminary adorn the house built on its site in the 1920s.

The most appropriate reminder of Louisa Barker is her own house which she built on Chestnut Street close to the White Seminary about 1851. She chose a design from a pattern book published by Andrew Jackson Downing, the most influential landscape architect of the mid-19th century. Downing shattered the popular opinion that architecture should display classical restraint. He was a dedicated romantic. He admired individualism and he prized freedom of expression. The great innovator believed that every house, whether set in a city or a village or in the country, should express the personality of its owner. "To find an original man living in an original house," he wrote, "is as satisfactory as to find an eagle's nest built on the top of a mountain crag."[2]

The owner of the charming Gothic cottage on Chestnut Street was not an original man. . . . She was an enterprising and original woman.

[1] Rev. A. D. Gridley, *History of the Town of Kirkland*, New York, 1874, p. 172.
[2] William H. Pierson, Jr., *American Buildings and Their Architecture*, Garden City, 1980, p. 270.

Louisa Barker lived at 1 Utica Street while she was preceptress at the Female Department of the Clinton Liberal Institute. The house was built in the late 1820s by Amaziah Stebbins, a merchant whose store was located on the site of today's Clinton Fire Department building.

Nathaniel Griffin built the big house in 1790 on land purchased from George Washington and George Clinton in the same year. The Griffin farm, a parcel of 316 acres, formed part of the 4,000-acre tract which Gen. Washington and Gov. Clinton had bought in 1784. At *that time George Washington was one of the largest land speculators in the United States.*

Nathaniel Griffin was born in Connecticut in 1761. He and his wife Parnell had nine children. They were members of the Congregational Church in Clinton. Mr. Griffin was *one of the original contributors to the Hamilton-Oneida Academy, and gave £ 4 in cash and six pounds payable in grain during Samuel Kirkland's 1793 drive for funds.*

The earliest known owner of the house on Bristol Road was W. Haven. The house was built before 1852.

The Gothic Revival rectory of St. James Church, Williams St., was built in 1867, two years after the completion of the church.

The house on West Park Row is one of the oldest in the village and probably dates from the 1830s. Othniel S. Williams, Jr. is believed to have had his law office in the building. At one time the Clinton Grammar School held classes on the second floor.

The original portion of today's Clinton House was built by Othniel Williams, Sr. about 1820. The Williams had moved from Killingworth (now Clinton), Connecticut to New York State six years earlier and had settled at first in Waterville. Mr. Williams was active in the affairs of Hamilton College and served as trustee 1827–32.

The Othniel Williams' son, Othniel, Jr. headed the second family which lived in the house. A graduate of Hamilton College (1831), trustee and later treasurer of the college, leading citizen, Oneida County Surrogate and Judge, Othniel Williams, Jr. supported many public improvements including the opening of Marvin, Chestnut and Williams Streets in Clinton. He remodeled the house in the 1870s.

The house at 25 Williams St., corner o
Chestnut St., was built about 1853, and i
was one of the first dwellings on the street
Williams St. was opened in 1850. The upstair
front room of the house, with bay windows
was added in the 1880s.

Mr. A. S. Taylor is believed to be the origina
owner of the house at 22 Marvin St. The
house was built probably in 1850, the year
when Marvin St. was opened.

Philip Hooker, the great Albany architect, designed the tower and steeple of Hamilton College Chapel. The exterior of the chapel stands
almost unchanged since its completion in
1827. The building is ''one of the best pro-
portioned and most charming of its kind in
America,'' according to the famous art critic
Edward W. Root of Clinton.[1]

Hooker was ''the best architect of that day
west of the seaboard,'' Root wrote.[2] His work
was ''extremely unsophisticated and also
overwhelmingly American . . . Had he had a
more thorough education, richer patrons and
more skillful workmen he might have sur-
passed Bulfinch as an architect.''[3]

The clock was given in 1877 by John
Wanamaker of Philadelphia.

[1]Philip C. Jessup, Elihu Root, p. 151.
[2]Edward W. Root, Philip Hooker, p. 19 ff.
[3]Ibid.

In the spring of 1792 Samuel Kirkland built this cottage on a portion of his land near the foot of today's College Hill, close to the site where he built his mansion a few years later. The cottage, 17' × 24', had a family room downstairs with fireplace, and three sleeping rooms upstairs. Kirkland, then a widower, lived here with his three younger children.

About 1836 the cottage was moved to a site on College St. at the west side of the Oriskany Creek, and for the next decades it had several owners, one of whom used it for a carpenter's shop.

The Clinton Rural Art Society, meeting Nov. 10, 1875 at the home of Prof. Oren Root, voted to buy the cottage and move it to a place on the campus just above the cemetery. The purchase price of $140 was contributed, and members pledged $50 to cover the moving costs.

In 1925, through the efforts of Elihu Root, the cottage was moved to its present location and restored. It was furnished in appropriate, contemporary style by Mrs. Edward W. Root.

Buttrick Hall was built about 1812 as a dining hall for the college. Students paid $1.50 per week for board.

In 1834 the building became the residence of Horatio Gates Buttrick, superintendent of grounds. The Buttricks with their daughters had moved to the college from Concord, Mass. In 1837 Nancy Buttrick, the eldest daughter, became the bride of Oren Root, later professor at the college. The Roots' son, Elihu, was born at the Buttricks' house in 1845.

Prof. Oren Root established his famous collection of minerals in a portion of the building, then called the "Cabinet," in 1850. Within a few years the rest of the building was converted to recitation rooms.

Buttrick Hall was restored to its original appearance in 1926.

North College was started in 1823. Hamilton was just 11 years old and two other dormitories had been built already. Funds were limited. After North's outside walls were built, the Hamilton trustees decided to spend money on a chapel rather than a third dormitory. They halted construction, sealed up the shell of North and diverted the unused materials to the erection of the chapel. Construction of North College was resumed in 1842 and the building was completed in 1846.

The dormitory did not have a central heating system. In addition to a sitting room, bedroom and closet, each of the students' quarters contained a stove and a coal room which could be filled from the hallway.

The original portion of the Root Homestead was built just before 1800. In 1803 it was enlarged under the supervision of the master carpenter Isaac Williams, and it became an inn. Later it was used as a boarding house for college students. Marcus Catlin, Professor of Mathematics at the college, bought the property for a residence in 1835.

Upon the death of Prof. Catlin in 1849, Oren Root was named Professor of Mathematics, and he bought the house. The property remained in the Root family until 1958.

Professor Oren Root is remembered with great affection and respect for his pioneering interest in mineralogy, astronomy, geology and many other branches of science. His enthusiasm for botany led him to start planting the trees, shrubs and gardens which give great charm to the Homestead today. He and his wife and sons began the transformation of a dump near the house into a lovely glen, which is cherished still.

In 1881 Prof. Root died and his son, Oren, Jr., succeeded to the chair of mathematics and the ownership of the Homestead.

Edward W. Root, son of Elihu Root and nephew of Oren Root, Jr., joined the Hamilton College faculty as Lecturer on Art in 1920–21. He and his wife Grace settled at the Homestead, and they continued to lavish loving care on the house and grounds in the family tradition. Edward W. Root became widely known as a collector and connoisseur of art. In 1958, after his death, the Homestead became the Root Art Center, an appropriate use for the home of one of the finest art critics in America.

The house which is today's Hamilton College Admissions Office was built in 1816 for Theodore Strong, Professor of Mathematics at the college. For many years the fine residence was the summer home of the distinguished statesman Elihu Root, Secretary of War under President McKinley, Secretary of State under President Theodore Roosevelt, winner of the 1912 Nobel Peace Prize, U.S. Senator, famous corporation lawyer. He graduated from Hamilton in 1864 as valedictorian of his class, served on the board of trustees of the college from 1883, and was chairman of the board from 1909 to 1937.

Rev. Samuel Kirkland built his mansion about 1797 on a portion of the large tract given him by the Iroquois in appreciation for his tireless efforts in their behalf. The land was just west of the Line of Property which, at that time, separated the Indians' lands from the whites'.

Rev. Kirkland studied at Dr. Wheelock's (later Dartmouth College) and graduated from the College of New Jersey (later Princeton University). He devoted his life and energies to missionary work among the Iroquois. In 1793 he founded Hamilton-Oneida Academy, an institution which tried to provide education for Indian boys as well as white boys. The Academy became the forerunner of Hamilton College.

The original owner of the house on Harding Rd. near Rte. 12B was John Kirkland who may have been a nephew of Rev. Samuel Kirkland. The house was built about 1812.

The large farmhouse at 1126 Lumbard Rd. was built about 1809 by Ralph S. Lumbard.

THOMAS DEAN

The main part of the Dean homestead was built about 1824. The original section, now adjacent to the house, was built in 1799, by the Indians, for the Deans.

In May, 1817, a bold young Quaker helped a band of Brothertown Indians build a large boat in the field close to his own house at today's Deansboro. The Quaker was the intrepid Thomas Dean. The boat was his weapon in a campaign to save the Brothertowns from the whites. He relied on the boat to carry a delegation on a voyage to the Indiana Territory, more than 1500 miles from his home. Like a determined statesman, he planned to lead his party to a council meeting of the great, friendly tribes of the west. The sanguine leader expected to wrest a treaty for a tract of land where the Indians of his own region might resettle in the wilderness of northern Indiana.

Thomas Dean was a huge young man. He was over six feet tall, broad-shouldered, and he had enormous strength and stamina. He had a good-humored face, with dark eyes under thick eyebrows, a strong

nose, firm mouth, and a square, jutting chin. At the time the boat was built he was 34 years old, and he had a wife, Mary, and five little children, all of them less than eight years old. He was a dedicated humanitarian, in the highest traditon of the Quakers. The approaching trip held no profit for him. His wife and children would probably have been much better off if he had stayed home to run the family grist mill and the farms, and let the Brothertowns succumb to the fate of so many others.

The bond of affection which tied the Indians to the Dean family had begun in the late 1790s, when the Society of Friends in New York City had sent Thomas' father to the New York wilderness as missionary to the Brothertowns. He was charged to "teach industry and morality"[1] to the defeated remnants of the New England tribes which had been given sanctuary by the Oneidas in the area near the Oriskany Creek.

The Indians had responded to his teaching with enthusiasm. They had become a model tribe, with a form of self-government patterned on the New England town meeting, and with laws against drunkenness and "frolicking"[2] on the Sabbath. Yet, white settlers were increasing in the area, and Thomas Dean saw danger ahead. The Brothertowns, like other Indians, were vulnerable to the diseases and the disastrous temptations of alcohol which seemed always to follow the white settlers. Thomas Dean was determined to avert a tragedy.

The boat, "Brothertown Enterprise," was finished by the latter part of May. It was a sturdy, open boat with a keel, oars and a sail. It had ample room for the party of eight with their baggage of blankets,

[1]Pomroy Jones, *Annals and Recollections of Oneida County, NY*, Rome, 1851, p. 264 ff.
[2]Ibid.

Deansboro

ropes, guns, spears and provisions. Built for a maximum burthen of six tons, the boat drew 21 inches of water.

When the boat was finished, Indians tugged and shoved it onto a heavy wagon. Oxen dragged the wagon to the launching point at Oneida Creek nearly 20 miles away, near Oneida Lake.

The party embarked at Oneida Creek. Thomas Dean, the only white, was captain, advisor, navigator, doctor, attorney. There were two chiefs with their wives, and two Indian men and a boy. They formed an elite group. In the words of an eye-witness who saw them in the west a few months later, all "possessed habits, manners and education indicative of the most refined civilization."[3]

On June 1st, according to Thomas Dean's journal, "at 3 a.m. . . . all set about getting ready to start. It was clear and very frosty; some ice. We drank some chocolate, ate some bread, all aboard, and went down the creek to the lake (Oneida). Set sail at 6 a.m."[4] The next day, after hiring a pilot to run the boat over Oswego Falls they "put out to sea"[5] on Lake Ontario. Ten days from home they reached the Niagara River, where they hired two wagons with ox teams to take their boat and baggage over the portage. Thomas Dean and some of the Indians walked on ahead to see the famous Niagara Falls. From Buffalo the trip down Lake Erie was so rough that several were seasick. The cook was "so sick he did not feel like doing anything and instead of having a warm breakfast . . . we were glad to take a piece of bread and raw pork, which relished very well. . . ."[6] They sailed beneath perpendicular rocks of Lake Erie's shore to the portage "over the land between the waters of Lake Erie and the Chautauqua Lake."[7]

From Chautauqua Lake they sailed down the Allegheny to Pittsburgh. Thomas Dean noted that the "town is well situated on the point, but it might be very considerably improved in beauty of appearance. Their manufactories are most interesting and surprising."[8] Moving west on "the pleasant current of the Ohio,"[9] they passed Cincinnati and lodged one night in

Louisville, "a handsome town of Kentucky."[10]

In early July the party reached the mouth of the Wabash at the corner of today's Kentucky, Indiana and Illinois. They were 1,546 miles from home, and the hardest part of the voyage still lay ahead. They had to push upstream on the Wabash. Sometimes "we had to get out and wade to find the channel, and shove the boat up."[11] They got chilled from the cold water, and several got sick. All recovered, and near Terre Haute they began the overland march into Indian territory, "fastened our boat near the fort to a stump, put the oars, poles, etc. into the blockhouse, and prepared to start. . . . They told us we could go through in three days if we had horses, it being 100 miles, so we concluded to take three days' provisions and get horses if we could."[12]

Dean walked the 100 miles to the meeting place where the fate of their mission was to be decided. Their magnificent effort foundered on the rocks of turmoil caused by the advancing settlements of the whites. The tribes were confused by events, and refused to promise land to the Brothertowns. The party of the "Brothertown Enterprise" started for home, disappointed and defeated.

Dean's spirits drooped. Homesickness gnawed at him. "This morning I awoke about 2 o'clock and put out the fire," he wrote about a week after the failure at the council meeting, "laid down again, went to sleep. I had a remarkable dream which agitated me very much. It brought to me the situation of my family, and the state of my affairs in which I left them. The imprudence of leaving home on such a journey without first settling all of my affairs; that they would lose greatly in case of my never returning to them again. The contents of my dream agitated me so that I could not eat much breakfast. . . ."[13]

The party returned to Terre Haute and headed east toward Detroit. When navigation became too difficult they sold the "Brothertown Enterprise." They walked many miles through the forests and then built a canoe for the last part of the remarkable trip. When they arrived in Detroit they booked passage for

Buffalo on board a schooner, took "a deck passage for $3 each, and lodged in the hold on the cable."[14]

In October, Thomas Dean returned to his family, in spite of the foreboding dream. Almost immediately, as if the failed mission had whetted his appetite for combat, he made plans for another attempt in behalf of the Indians. For the next ten years he threw his enormous energy into a frenzy of travel. He argued in the New York State Legislature; he entreated the U.S. Congress; he pleaded with the western chiefs, "going to the city of Washington six times, to Green Bay, Wis. four times, and sundry other journeys, in all about 20,000 miles."[15]

Dean triumphed in 1828. As Quaker missionary, and New York State agent for the Brothertown Indians he made a treaty with the U.S. Government. The treaty allowed the Brothertowns together with their neighbors, the Stockbridge Indians, and some of the Oneidas to move to a large tract of 23,000 acres on the east shore of Lake Winnebago, near Green Bay, Wisconsin.

In 1831 the first group of Indians sold their farms near the Oriskany Creek and started for their new homes in Wisconsin. They traveled west via Erie Canal packet and lake steamer. Over the next few years the orderly emigration continued until the total of re-settled Indians reached 2400.

As the emigration of the Brothertowns proceeded, the Chenango Canal transformed the little settlement near the Deans' home. Gangs of Irish laborers dug the 40-foot wide ditch through the hamlet. The mile-long Section 13 passed through the "land of Thomas Dean, Esq., and . . . from the foot of the bank opposite his house, along the border of the valley."[16]

The canal opened in 1837 and made the settlement flourish as never before. Two warehouses were built. Merchants built two stores, with platforms on the back at water level and

---

[3]Thomas Dean, *A Voyage to Indiana in 1817*, published by John Candee Dean 1918, reprinted by Town of Marshall Bicentennial Committee, 1976, p. 275 ff.
[4-15]Ibid.
[16]F. C. Soule, comp., *The Chenango Canal*, Canal Society of New York State, Syracuse, 1970. p. 6.

pulleys and ropes which hoisted goods from the canal boats to the two upper floors. Two taverns kept busy, and a provision store prospered. The Deans' farms, grist mills and brewery did more business than ever. The canal brought so much activity that the post-office of the Town of Marshall moved from nearby Hanover to a site near the Deans' house. Thomas Dean, became postmaster of Deansville.

Thomas Dean died in 1844, a short time before the last of the Brothertowns left Deansville. During the years that he said affectionate farewells to the groups of departing Indians, many tragic migrations of Indians were taking place all over the eastern United States. As Indian lands were liquidated, thousands of Indian families trudged a bitter "Trail of Tears." Rapacious land-grabbers drove them from their homes, and unscrupulous officials abandoned them to their dismal fate.

Unlike the majority of United States Indians, the Brothertowns were extremely lucky. They traveled a pleasant path from their farms near the Oriskany Creek to their new homes in Wisconsin. Their amazingly agreeable migration was a tribute to a compassionate and persevering friend.

The Peck homestead dates from the mid-19th century. Ever since the house was built the property has been noted for its fine gardens. According to a family tradition, the original owners of the house used to invite village residents to walk in the gardens and pick the flowers.

This mid-19th century house is an example of the octagon style originated by Orson Fowler. A Home For All, *the architectural guide published by Fowler in 1854, held that the octagon style was cheaper, roomier, more convenient than other styles, and since there were no dark corners it was more light and cheerful. "A given length of octagon wall will enclose one fifth more space than the same length of wall in a square shape," he wrote.*

Oriskany Falls Stone Church was built in 1834–1845 by the Congregational Society. The limestone blocks are held together in mortar made of crushed limestone, cement and wood ashes. In 1888 the new steeple was added, to replace the earlier one which was blown off in a storm.

The original portion of the Isaac Allen farmhouse was built in 1797. The house stands on College Street on the outskirts of Oriskany Falls. It was enlarged in 1832.

In 1876 the farmhouse became part of the property owned by Sidney Putnam, the prosperous cattle dealer, farmer and landowner who opened the quarry at Oriskany Falls in the early 1850s. He built a loading dock on the canal and shipped limestone in his own barges. Because of its excellent quality for building, the stone was in demand as far south as Binghamton.

The farmhouse is owned today by descendants of Sidney Putnam.

Oriskany Falls

The original portion of the house on Main St. owned today by Miss Anne Douglass was built probably between 1835 and 1840 for Chatfield Alcott, believed to be an uncle of Louisa May Alcott.

Miss Douglass' grandfather, James Alancthon Douglass, once principal of the Oriskany Falls High School, bought a warehouse on the Chenango Canal in 1867. For many years the warehouse, on Main St. near the center of the village, was a landmark of Oriskany Falls. In partnership with E. A. Hamlin, James Douglass ran a lumber yard and produce business in connection with the warehouse.

The prosperous hop farms in the vicinity probably accounted for a large part of the lumber business. Not only were the farmhouses and barns built on a generous scale, but many farmers built large additions to their houses to provide sleeping quarters for hop pickers at harvest time. Hundreds of thousands of pounds of hops were picked in Oneida and Madison counties each year and shipped by canal. The business was very profitable, particularly in the 1850s and 1860s. Many big farmhouses with cupolas were built with "hop money."

In addition to hops, the firm of Douglass & Hamlin shipped apples, grain and dairy products via canal. Boats which docked at the warehouse brought coal, plaster and salt, as well as lumber, lath and shingles.

Today, few vestiges of the Chenango Canal's 116 locks remain. One of the best-preserved lock walls may be seen on the north side of Valley Rd., about midway between Oriskany Falls and Solsville.

From Solsville, the Chenango Canal ran south through Bouckville. Several buildings within the village of Bouckville are included in the authors' previous book, Cherry Valley Country.

Eaton

Deacon William McCullis' home, Eaton Rd. north of today's Eaton, was built between 1800 and 1807. McCullis bought the land in 1797 or 1798 in the township of Eaton, one of the Chenango Twenty Towns, first opened for settlement by the Clinton treaty of 1788.

The Federal style house with stepped, Dutch gable ends has four chimneys which serve eight fireplaces. The date 1807 is painted on the inside of the attic roof.

The house of the Joseph Morse family was built in 1802. It stands on today's Route 26, in Eaton.

THE MORSE FAMILY

"Hereditary physical strength, great mental activity and unconquerable perseverance, the dominant traits of this family . . . eminently fitted them to assume the duties and bear the hardships of building up the new country and to nurture the institutions of civilized life."[1]

The country was very new indeed when Joseph and Hezekiah Morse and their families arrived from Sherburne, Mass. in 1796 to settle beside the roaring stream at today's Eaton village. Joseph Morse built his first dwelling of hand-hewn boards beside the well-traveled Indian trail which led from Oneida Castle to the Susquehanna. Indians camped here frequently, and the Morse children had Indian children for playmates. The Morse latchstring was always out, and on many mornings the family found Indians sleeping with their feet to the Morse fireplace.

Unlike the modern industrialist who can choose between steam, oil, electric or atomic power to run his plant, Joseph Morse had only one option. Waterpower. Eaton Brook, rushing close to his house, provided magnificent power, and he harnessed it early. Soon after the family settled, he built a dam and a grist mill. A sawmill, plaster mill, foundries, a factory for making wooden water pipes and suction pumps, a distillery followed shortly. Within a few years Joseph Morse helped start a woolen manufacturing company. The companies and mills prospered, and Joseph Morse became the largest landowner in the vicinity and the unofficial banker as well. He built a big stone house, owned rich farmlands, said to be the best in the area. He raised cattle and planted an orchard which was remembered for many years. He worked hard to improve the roads from Eaton to market, and he helped finance and build the Hamilton-Skaneateles turnpike which provided a vital access to market for his own and his neighbors' products.

In 1814 Joseph's daughter Eunice was married in the lovely stone house. The bridegroom was Dr. James Pratt, the pioneer physician and schoolteacher of the village. The wedding was one of the county's most brilliant social events, and the guest list sparkled with the names of the leading families for miles around. The John Lincklaens came from Cazenovia. The Morris family traveled from Morris Flats (now Morrisville). The Peter Smiths and their two sons Gerrit and Peter Skenandoa Smith came from Peterboro in their coach with driver and footmen. The Lelands drove up from their home on Leland Lakes, "the still beautiful Waistill Leland and her seven sons and three daughters . . . escorted along the turnpike by a party of Indian friends dressed in beautiful beaded buckskins, mounted on high-stepping horses."[2]

The wedding took place in the front parlor at 2:30 in the afternoon. Eunice came down the stairs with her father. She wore a simple white wedding gown and carried a bouquet of wildflowers. During the reception guests drank punch which was served from large silver bowls. A string trio of violin, cello and bass viol played, and couples danced the minuet, gavotte and shotchische beneath lighted crystal chandeliers.

Five years after Eunice's wedding, Joseph Morse and his son Alpheus rode to Boston on horseback, driving a herd of their own and their neighbors' cattle to market. On the way, Joseph became ill. After taking the herd to Boston, he went to his old village of Sherburne, trying in vain to recuperate. He died in Sherburne.

Joseph's death did not stop the torrent of family activities. His sons, nephews and grandsons carried on the enterprises already established, and added more. They continued to be innovators. In 1835, Elijah, a nephew, bought a mechanical hay rake, the first ever seen in the county. Later he helped acquire the county's first mechanical mowing machine, buying it from the inventor for $65. A grandson, George, introduced the cheese factory and creamery system.

The family is remembered best for its largest enterprise, Wood, Tabor & Morse. Among the company's products were portable steam engines which were manufactured for many years beginning in 1855. The Wood, Tabor & Morse engines were famous in the United States and abroad. The Eaton company shipped many of them to the west where they operated threshing machines. They became a well-known sight on the prairies. The little, self-propelled engines were driven along the roads from farm to farm, "pulling the threshing machine . . . and bunkhouse behind."[3]

The steam engines were shipped from Pecksport, just a few miles down the hill from Eaton. The hamlet and loading dock occupied a site on the Chenango Canal about midway between Hamilton and Bouckville.

Although the steam engines were manufactured profitably, they puttputted a message of doom to Eaton as they chugged along the roads. Soon steam engines destroyed waterpowered mills everywhere.

In 1869, before steam power put most of the water wheels out of business, a lady visited some relatives in Eaton and she wrote home, "Looking down from . . . the cottage and following the course downstream as it flows eastward from Eaton Brook Reservoir one can see the many mills and factories there which are powered by the racing water."[4] Within a few years, steam power lured industry away from the finest rushing streams. The Morse empire, based on water power, succumbed.

[1]Mrs. Hammond, *Morse Biography*.

[2]Hart, *Last Gathering*, p.12

[3]Houghton, p. 54.

[4]Hart, *Lebanon Hill Journal*.

The brick house built by Samuel Sherman about 1800 is probably the oldest dwelling in the village of Eaton.

The schoolhouse, now a residence, was built on today's River Rd. about 1850, south of the Horace Sherrill farm.

U.S. common schools improved steadily in the period 1830–50. The New York State budget for common schools rose to $220,000, one-half being raised by local taxes. All students paid tuition until 1849. In that year the Legislature voted to provide free education for all. The act was repealed two years later, but a Free School act was passed once more in 1867.

In 1828 Horace Sherrill and his wife Lydia built the stone house on River Road less than three miles south of the village of Eaton.

Lydia's father, David Shapley was the first member of the family to own the land where the house is built. A former elder in the Shaker colony at New Lebanon, David Shapley with his brother and a friend and three Shaker ladies left the colony in 1790. The couples married and settled in the township of Lebanon, then a wilderness near the Chenango River. No roads existed in Madison County at that time, and blazed trees marked the only trails between settlers' homes.

The stone farmhouse on River Rd. was built in 1832. Stone was quarried in the present woods back of the house.

The Chenango Canal feeder from Kingsley Brook (now Lebanon) Reservoir ran in front of the house. Feeders at the summit level were an extremely important factor in the design of the Chenango Canal. The seven feeders held back a total of over 11 billion gallons of water.

The cobblestone house on River Rd., less than a mile south of Earlville, was built for the Crandall family between 1835 and 1840. It is believed to be the only cobblestone house in Chenango County.

"Maple Hill," the large stone house on River Rd. was built about 1836 during a temporary lull in the construction of the Chenango Canal. A large gang of stonemasons, lock builders, were laid off because of a lack of materials, and they were hired to build the house. They worked very fast. According to one legend the house was built in 14 days.

The house at 40 Broad St. was built for Mr. and Mrs. Esek Steere 1816–25.

ESEK STEERE

As a great wave surges, pounds, then fans out in gentle endings, so the New England emigrations of the end of the 18th century flung thousands of settlers into the New York wilderness, and a milder momentum propelled the pioneers' descendants to their own resting places. In 1796 Esek Steere, Sr. marched hundreds of miles from Rhode Island to Otsego County and settled there; in 1816 his son, Esek, Jr., moved about 30 miles farther west to settle at today's Hamilton.

Esek Steere with Mary, his bride, moved to Hamilton in the year when the little settlement started by the vigorous Payne brothers became an incorporated village. Residents dropped the name "Payne's Settlement," and re-named the place in honor of Alexander Hamilton.

Already the little settlement had several stores, a tavern, some fine residences, a doctor and at least one lawyer. As if to celebrate their new status as a village, residents opened an Academy. The village public school met on the ground floor of the building at the corner of Broad and Pleasant Streets, while the second floor held a private secondary school, Hamilton Academy.

In 1816, Esek Steere, the newcomer, went into partnership with Joseph Cowell, one of the pioneer merchants of the village. The Cowell store stood at the corner of Broad and Lebanon Streets, close to Elisha Payne's tavern. It was a prime location. Heavy farm wagons and light, horse-drawn chaises passed frequently, and scores of pedestrians picked their way along the muddy streets. The partners tore down the Cowell store, giving the timbers to a neighbor who used them for building his barn, and they built a new brick building. They opened a mercantile business here, and it flourished.

In 1819, residents of Hamilton elected Esek Steere a village trustee. Five years later he became a trustee of the Hamilton Academy. This was a time when a significant change transformed the noted institution. The trustees had taken a giant step. In order to persuade the Baptist Theological Seminary to locate in Hamilton rather than in one of the other villages which sought its presence, they had built a third story on the Academy and offered it to the seminary for classrooms. When Esek Steere became trustee of Hamilton Academy, the seminary classes were already meeting here, and they continued to meet in the building for three more years until the seminary moved to its new home on the Payne property and became Madison University, later Colgate University.

Soon after the Steeres arrived in Hamilton they built the lovely house which stands today at 40 Broad Street. Esek's chief interests were comfortably close by. His business partner, Joseph Cowell, lived in a fine house next door. The brick store was only a block away. The Hamilton Academy stood just across Pleasant Street.

The Steeres had no children, but they liked young people and they were notably kind and generous to the pupils at the Academy and, later, to those at the University. Esek is remembered as "befriending many students by timely gifts and giving homes to those who were dependent."[1]

Construction of the Chenango Canal began in 1834. It promised future prosperity for a mercantile business, but the first years of its construction brought some perplexity for the store owner. The construction workers lived almost in the Steere's back yard. Villages of shanties called "Paddy Towns" clustered in the canal villages during its construction years. Hamilton's village of shanties ran along today's Maple Avenue, the little street on the bank of the canal, parallel to Broad. Residents of Hamilton called the street of shanties "Canada," for the people who lived there seemed to belong to a faraway place.

Wild Irish immigrants, "Bog Men," or "Bog Trotters," did most of the manual labor on the Chenango Canal. They wielded the shovels, guided the horse-drawn dirt scrapers, pushed the wheelbarrows, dumped the tip-carts. They drank large amounts of grog, and they fought fiercely and frequently. The men were very poor, driven to the U.S. by poverty in their native land.

Regular wages of $11 to $15 a month seemed princely in comparison to those abroad. With the opening of the canal the shanties began to disappear. Laborers moved on. Many went West. Others joined construction gangs on the new railroads. When the Irish left, tranquillity returned to Hamilton, and residents felt that the canal-side street belonged to them again.

The opening of the canal stirred a surge of activity in Hamilton, New stores opened. Prospering residents built new houses. A new tavern started in business. Esek Steere started a hardware store in partnership with two men of the village, John Foote and E. W. Foote. The partners took down the brick store at Broad and Lebanon Streets and built a much larger one on the same prime site. They needed more space than before, to stow and display the heavy articles which could be floated to Hamilton on the canal boats . . . stoves, sheet lead, bolt copper, pig iron, sheet iron, bar lead, nails, saws, cauldron kettles and the like. This was one of the first hardware stores in the Chenango Valley, and it prospered for decades. In Esek Steere's day, the sign over the door read "E. Steere & Co." Later, when he withdrew from the business, it carried the name "Foote & Gaskill."

Esek Steere died in 1846. He was 61 years old. His wife Mary lived until after the Civil War.

During his 30 years in Hamilton, Esek Steere built a reputation as a valuable and generous citizen. Perhaps the greatest tribute to his kindly personality came many years after he and his wife had died. In 1898 an alumnus of Hamilton Academy recorded his memories of the village when he arrived to enter the Academy as a student in 1825. He tried to remember how Broad Street looked when he boarded at Deacon Olmstead's south of the village, and walked to his classes at the Academy each day. Although the old maps show that many handsome houses lined Broad Street in the 1820s, the alumnus could remember only one, and that one belonged to Esek Steere.

---

[1]Steere Genealogy.

The original portion of the house at 42 Broad St. was built about 1808 for Joseph Cowell, one of the first merchants in Payne's Settlement, today's Hamilton. Cowell opened a store in the settlement about 1800.

George Williams, printer and publisher, built the original portion of the Federal style house at 60 Broad St. about 1835.

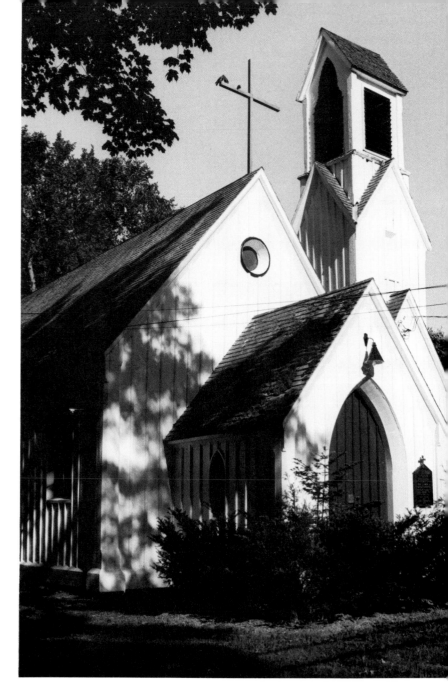

St. Thomas Episcopal Church, 14 Madison St., was designed by Richard Upjohn and built in 1846–47. The cornerstone was laid by Bishop DeLancey, bishop of Western New York.

The architect Richard Upjohn was largely responsible for introducing the Gothic Revival style in American church building. The style was particularly popular among Episcopal and Roman Catholic parishes.

Nelson Wilcox, a lumber dealer, was the original owner of the pre-1850, Gothic Revival house at 38 Maple Ave. The street parallels the route of the Chenango Canal and was known once as Canal St.

Willard Welton is believed to be the original
owner of the house at 25 Payne St. A Connecticut native, Yale graduate, Mr. Welton began
the practice of law in Sherburne, NY in 1809.
He moved to Madison ten years later and settled finally in Hamilton in 1835. The house
dates from the 1830s.

The Federal style brick house near the eastern end of Payne St. is believed to have been built about 1830.

The house at 27 Payne St. was built in the early 1830s, probably for Deacon Charles C. Payne, a son of Elisha Payne.

The house at 35 Payne St. was built for the Elisha Payne family about 1800. Elisha Payne, with his wife Polly and four children, came from Lebanon, Ct. to today's Hamilton in 1795. He joined his brother Samuel Payne who, with his wife Betsy, had settled here in the previous year.

A very energetic man, Elisha Payne became the pioneer developer of the settlement known first as "Payne's Hollow," then "Payne's Settlement," and incorporated as Hamilton in 1816. The first town meeting was held at Elisha Payne's house. He built a tavern at the corner of Broad and Lebanon Sts. in 1802. He served as associate county judge and justice of the peace. He was a founder of the Baptist Education Society of the State of New York, the organization which started the institution which became Colgate University.

The Crocker family built the house at 30 Broad St. about 1832. An office building with apartments today, the Federal style brick house retains its original exterior beauty through careful restorations in 1955 and 1976.

The house faces the village green which was a swampy place before the construction of the Chenango Canal. The area was filled in and the charming green was created with many cartloads of earth dug from the canal route.

West Hall, completed in 1827, was the first dormitory at Hamilton Literary and Theological Institution, later Colgate University. The Institution, founded by the Baptist Education Society of the State of New York, trained young men for the ministry and for missionary work. From 1839 the Institution accepted liberal arts students as well as those training for the ministry. The Institution was chartered as Madison University in 1846, and became Colgate University in 1890.

The original building contained a chapel in part of the third and fourth stories, as well as a lecture room, study rooms, library, and sleeping rooms for about 70 students. Some students earned part of their expenses by preaching at nearby churches and teaching in district schools. Students in the dormitory provided their own wood and coal for stoves, brought their own tallow candles or whale-oil lamps, and bathed in a bath house supplied by spring water piped down the hill.

The house on today's Route 12B, a little south of the Colgate campus, was built before 1810 for the farmer Jonathan Olmstead, deacon of the Baptist church in the village.

The house is cherished today as the site of the September, 1817 meeting at which six clergymen and seven laymen formed the Baptist Educational Society of the State of New York. The Society started Hamilton Theological and Literary Institution, forerunner of Colgate University. The Institution was chartered in 1819.

The big stone house on Preston Hill Rd. is thought to have been built in 1793. For many years it was known as the Smith homestead. The Adon Smiths lived in the house from 1830, and were followed by their son J. D. F. Smith and his family. Adon N. Smith, a grandson, was born here in 1854.

The Page family homestead, 52 East Main St., was built in 1856.

THE PAGE FAMILY

In the 1790s, settlers who came looking for land, called it The Forks, or sometimes, Madison Forks. The little settlement was a cluster of cabins near the big old Indian camping ground between two branches of the Chenango River. Soon, pioneers built a tavern, a distillery, a tannery, an ashery, some blacksmith shops and several more little dwellings. About 1804, the residents painted all these buildings red. For a while the place was called Red City. Later, perhaps because the paint faded, it became The Forks again.

When news of the building of the canal raced down the Chenango Valley, communities celebrated with parades and fireworks and parties. The Forks made an even more lasting gesture. It adopted a new name to match the big changes to come. The postmaster chose the name Earlville. It honored his close friend, the canal commissioner Jonas Earll.

The canal altered the little settlement in size and scope as well as in name. Earlville grew and prospered. More people came here to live, and several men in the town built new businesses. One of the foremost of these was George M. Page. He created Earlville's successful storage and forwarding business.

George Page started his canal-side enterprise with one warehouse. Around 1850 he built a new storehouse on the bank of the canal, about one canal boat's length south of the canal on East Main Street. In 1856 he built the big white house on East Main Street, close to the canal and the warehouse. For many years the Page warehouses gave the farmers of the rich and fertile land around Earlville a chance to ship out their grain, butter, cheese, hops, while canal boats brought them coal, salt and farm supplies.

George Page was a busy man, and he was a leader in the community. He became postmaster of the village. He ran an ashery back of his house. A driver remembered as Bill Brown used to go around town with an ash wagon drawn by a team of mules. He collected ashes and unloaded at the ashery, where the ashes could be leached, boiled, and then shipped on the canal as potash.

The ashery was just a sideline. George Page's main business was the warehouse and adjoining store, and the shipping on the canal. The business flourished. He bought three canal boats himself, and he opened a branch warehouse in Norwich.

The storage and forwarding enterprise was a family business. After George M. Page's death his son Caleb tried to carry it along. "C. S. Page, Storage, Forwarding & Commission Merchant," states an old advertisement, "orders for all kinds of country produce promptly filled." Forces beyond his control doomed Caleb Page to failure. The railroads pushed through the Chenango Valley in the 1870s, and the canal-side business collapsed. Caleb Page became a station agent for the New York Oswego, & Midland Railroad, at Earlville.

Within a few years a fire destroyed the Page warehouses, but the big white house survived. Three generations of the Page family lived here. Caleb's daughter Charlotte was born here in 1868. Like a soft west wind blowing over the nearby fragment of the abandoned canal, gracious memories hover around the lovely house. A friend who visited the Pages when Charlotte was a girl recalled, "The grounds looked very pretty — the tennis set, the croquet set, the hammocks, tent and rustic seats."[1]

Charlotte Page became a prominent educator and an early advocate of women's rights. "We were a moderately prosperous family,"[2] she wrote once. It was a comfortable prosperity created by an enterprising businessman and the Chenango Canal.

[1]Donna Burns, *The Letters of Charlotte Browning Page* in "The Courier," Syracuse University Library Associates, vol. XVII, No. 1, Spring, 1980, p. 15.
[2]Ibid, p. 15.

One stone arch of the original two-arch
aqueduct survives on Billings Rd. off Rte. 12D
south of Earlville. The aqueduct crossed the
east branch of the Chenango River. A total of
ten aqueducts carried the canal across the
Chenango River and other streams.

The red brick house at 22 South Main St. was built about 1815.

The cellar of the frame farmhouse on Preston Hill Rd. north of Earlville contains a section of mortar which is inscribed "N. L. Richards built September, 1849."

Poolville

The Greek Revival house in Poolville was built
for Richard Berry about 1830. At that time,
Poolville was a busy village, with a tannery, a
boot and shoe manufactory, a woolen mill, a
shop which made factory machinery, a hotel
and more than one tavern.

Sherburne

The Joshua Pratt home, Sherburne, was built in 1835. Mr. Pratt, a Connecticut native, settled at Sherburne in 1800 and started a mercantile business. In 1833 his sons Joshua and Walstein took over the business. Later the firm became Pratt & Rexford, and ran a store on Main St. and a warehouse on the Chenango Canal. (Several Sherburne houses are included in the authors' previous book, Cherry Valley Country.*)*

North Norwich

In 1812 Albert German built the tavern which is a residence today in North Norwich. The German tavern occupied an excellent location on the Cayuga Turnpike which connected Ithaca and Cooperstown. Street names in North Norwich, "West Cayuga" and "East Cayuga" preserve the memory of the old thoroughfare.

The home of Hosea and Sophia Dimmick, 21 Mechanic St., was built in 1839.

HOSEA DIMMICK

The boat captain probably had the most glamorous job on the canal, and the lock tender had one of the least. Yet the captains could not move their boats along the waterway unless the faithful lock tenders opened and closed the gates. The tenders were as indispensable as the locks themselves.

Hosea Dimmick was lock tender for Lock #93 at the end of Lock Street, Norwich. His salary was $20 a month. His superior, the Division Superintendent received $58 a month and reported to the New York State Canal Commissioner for the Middle Division of the Erie Canal, and the State Engineer.

Lock #93 raised or lowered the water level only six feet. At the lock, the towing team pulled the canal boat into the lock chamber while the steersman guided it carefully so that it would not graze against the sides. When the boat was completely contained in the lock, Hosea leaned on the long, heavy wooden balance beam and closed the massive wooden gates which swung to a V shape behind the boat. He walked forward along the side of the lock and opened the sluices in the bottom of the front gate. Water flooded into the lock from the upper level and raised the boat. When the water in the lock reached the level of the canal ahead, Hosea opened the upper gate and the boat proceeded.

Hosea's job demanded physical strength and it charged him to be completely honest. In addition to working the balance beams and the sluice gates, Hosea checked the captains' papers. Since Norwich was not a collector's office, he checked the bills of lading and the record of tolls paid at collector's offices in Utica, Hamilton, Oxford or Binghamton.

In 1839 Hosea Dimmick built the house which stands today at 21 Mechanic Street. He was 32 and his wife, Sophia, was 29. During the early years the house overflowed with sadness, for two daughters and one son were born, and none of them lived to be more than six years old.

Since the house was close to Hosea's work, he could probably hear a captain's horn if he didn't happen to be in his little shack near the lock. The dignified, pretty house stood in an industrial neighborhood, for like a magnet, the lock drew manufactures to its vicinity. A tannery at the east end of Lock Street dumped its waste water into the canal. Just below the lock, on the east bank, a planing mill screeched. In front of Hosea's house on Mechanic Street the water from the canal flowed into a "basin" which reached all the way to North Main (today's North Broad Street). Log rafts floated from the canal to the lumber yard at the west end of the shallow basin.

The foremost industry near Hosea's lock and house was the Maydole Hammer Factory. In 1845 David Maydole had bought land near the canal and had arranged with the state to buy water power from the waste weir at Lock #93. He made edge tools and carriage springs and he perfected the high-quality adz-eye hammer for which the company became famous. After an early fire he rebuilt at the same site and continued to manufacture hammers, using the water power from the weir. The Maydole factory employed 50 to 60 people. After 1860 the factory changed to steam power, and the work force sometimes reached 100.

In addition to hammers, the Maydole factory made ice skates. Mr. Maydole brought much pleasure to the residents of Norwich, for, in order to promote the sale of skates, he cleared the snow from the canal and basin near the factory, and he groomed the ice to a fine sheet. The sport became very popular. Soon, when weather conditions permitted, skating parties skimmed from Norwich to Oxford and from Norwich to Sherburne.

Winter Sundays were probably the pleasantest times for Hosea and Sophia. The sawmill and hammer factory were quiet, and there was no activity at the tannery. They could watch the skaters wheel and glide, turn and slide on the smooth ice in front of their house. On those days a lock tender's life was a very cheery one.

The house at 140 North Broad St. was built
for Mr. and Mrs. Andrus Pellet in the early
1870s. Mrs. Pellet, the former Mary Ann
Gorton, grew up in North Norwich where her
father had a ship chandler's store beside the
Chenango Canal.

Edward Child was the original owner of the house at 133 North Broad St., built in 1849 and 50. Mr. Child, Sheriff of Chenango County, was a carpenter and joiner by trade.

The Greek Revival house at 136 North Broad St., a Norwich landmark, was built in 1838 for Dr. Blinn Harris, a son of the pioneer Capt. John Harris.

Capt. Harris brought his wife Tamar and their six children to the site of today's Norwich in 1790. The family rode in a sleigh drawn by a big team of horses. Two yokes of oxen dragged a long sled which held the household goods and supplies.

Capt. Harris, a surveyor, laid out roads and townships in Chenango County. He laid out today's Broad St., Norwich, in the early 1790s.

DeCalvus Rogers was the original owner of the house at 123 North Broad St., completed in 1848. Mr. Rogers, born in Groton, N.Y., ran a successful grocery store in Norwich.

Dr. Harvey Harris, a beloved physician of Norwich, built the house at 110 North Broad St. in 1838. Dr. Harris, son of Capt. John Harris, was born in Norwich, 1795. He studied medicine under Dr. Mitchell, Norwich, and graduated from Albany Medical College in 1816.

Dr. Harris built a small office on the south border of his property. He maintained a large garden and grew many herbs for healing. He built a reputation for devoted service. It was said of Dr. Harris that he never turned down a request for help, no matter how bad the weather nor how far away the sick person lived.

Hiram Weller, a Norwich hardware merchant, was probably the original owner of the house at 103 North Broad St., built in 1837. The Federal style dwelling has a balanced facade with four two-story Ionic pilasters.

The Federal style house at 8C West Cortland St. dates from the 1820s, when the occupants of the residence at the corner of Broad and Cortland Sts. built the house in their "side yard" and gave it to their daughter as a wedding present.

The house is noteworthy for its four two-story pilasters and its very beautiful Georgian doorway with fan window.

Chenango County Courthouse was completed
in 1838. Contemporary newspapers called it
the most elegant courthouse in the state. The
cupola was adorned with a life-sized figure of
Justice, carved from the mast of a sailing ship.
It was purchased in New York City and
brought to Norwich by canal boat. Today the
original Justice is on display inside the courthouse, while a replica stands on the cupola.

Jonathan Wells was the original owner of the
Italian villa at 42 South Broad St. The house
was built in 1850.

The house at 79 South Broad St., probably the oldest house in Norwich, was built in 1797 for Dr. Jonathan Johnson, the first practicing doctor in Chenango County. Dr. Johnson, a native of Canterbury, Ct., studied medicine at Pomfret, Ct. He arrived in Norwich in 1794, on horseback.

The Balcomb homestead stands on Rte. 12 at Turner St., south of Norwich. The stone house was built in 1825. According to legends, it was a stop on the Underground Railroad. Escaped slaves were hidden here while making their way north to Canada.

The big grist mill on West Main St. at the Canasawacta Creek was built for Col. William G. Guernsey in 1836.

The large white frame house at 71 South Broad St. dates from the 1840s.

"HONEST JOHN" TRACY

John Tracy moved slowly toward prominence. Like a passenger on a canal packet he had ample time to observe the scene through which he passed, and he watched most of the disappointments and successes which stirred the lovely village of Oxford in the 19th century.

In 1805 John Tracy had moved to Oxford from the little Chenango County settlement of Columbus. It was the year when axemen felled the last massive trees to make Oxford the western terminus of the new Catskill-Susquehanna Turnpike. The turnpike gave the village a bumpy thread of communication with the older, big communities of the east. Weekly stage and mail service began. A stage left Catskill on Sunday morning and reached Oxford Tuesday, bringing passengers and news from New York City.

When John Tracy arrived, Oxford was a half-shire town, dividing the prestige and legal business of a county seat with Norwich. Two lawyers had already opened offices. John Tracy began to read law in the office of the affable Stephen O. Runyan, on Washington Square. The law student needed money to cover his expenses, and he became deputy to the county clerk, and served as village postmaster. When he passed his Bar exams in 1808, the Runyan firm became Runyan and Tracy.

More lawyers had opened their offices on the square. Henry Van Der Lyn started his practice. James Clapp and William Price came by wagon from New York City, bringing their chairs, desks and law library, and they put out their sign at a small building which they rented from a milliner.

More residents came from New York City. The most fascinating and colorful of these were the Benjamin Butlers. Mr. and Mrs. Butler and their daughters arrived in 1806, and they moved into a house on the northwest corner of the square. Mr. Butler had started his career in the shipping business in Massachusetts, had succeeded handsomely, and moved to New York City where, in the early 1800s, he was a prosperous broker. He stood out among the residents of Oxford for his elegance. An historian remembered him as "a dark-skinned man . . . of a fine, commanding presence."[1] The family brought a life style which differed sharply from that of the other people in Oxford. The same historian recalled that "Mr. Butler owned a colored boy as part of his personal property, as in those days New York was a slave state."[2]

In 1814 the Legislature voted to make Norwich the county seat of Chenango County. Norwich celebrated; Oxford mourned. Some of the lawyers prepared to move away, but John Tracy stayed in his adopted village. He married a Connecticut girl and they had twin daughters, and a son. The son died quite young, drowned in a skating accident on the Chenango River.

John Tracy built up a lucrative law business. He was rather an aloof man. His mouth was set in a straight line, as if to announce that he planned no deviation from any path he had set for himself. He earned the nickname "Honest John," for people trusted his integrity. He became Surrogate, First Judge and then Circuit Judge. In 1820 he won election to the New York State Assembly, and was re-elected for several terms.

Each term, as he left for the Assembly at Albany, he knew that his district needed help desperately. The fertile valley of the Chenango River yielded generous crops, but the farms furnished meager profits. All the farmers along the Erie Canal could float their grain and cheese and butter to New York City at a fraction of the price a Chenango Valley farmer must pay, whether he chose the rugged turnpike to Catskill or the equally rough and tedious road to the Erie at Utica. The dearth of transportation strangled the growth of trade and commerce and choked the flow of produce in the Chenango Valley, while new towns and cities sprang up along the Erie and the old ones prospered, and the banks of the canal filled with mercantile establishments and rich warehouses.

Tracy's constituents demanded some of the prosperity which, like a mirage, hovered just out of their reach. Their demands were just. Votes and taxes from the southern counties had helped build the Erie, and Governor Clinton had promised, in return for their support, that someday a lateral canal would tap the Erie and bring some of its riches down the valley.

Canal advocates believed the Chenango Valley was admirably suitable terrain. Many experts surveyed the route and assured them it was extremely practical. One of these, Benjamin Wright, Chief Engineer of the Erie, said, "The Valley of the Chenango from the town of Madison presents a formation of ground most extraordinarily favorable for easy excavation of a canal; so much so that I do not think the whole state of New York can present a similar distance where nature has given a formation more favorable for such a work, and more easily and cheaply executed."[3]

In the Assembly, John Tracy helped the committees in their struggle for the Chenango Canal. In spite of the success of the Erie, those who controlled the Legislature stood firmly against any more "internal improvements." Committees for the canal demanded, pleaded, persuaded, exhorted. Their opponents refused to yield. Anti-canal forces challenged the estimates of cost of construction; they disparaged the proposed route; they worried about the water supply; they doubted the possible revenues; they dreaded lawsuits which might arise when streams were diverted. The wrangling went on for nearly seven years, but it did not divert the canal advocates from their goal.

The year 1832 brought victory. William L. Marcy won the election for governor, with John Tracy as his running-mate. In his January address Governor Marcy said that the proposed canal should not be decided on

---

[1] Henry J. Galpin, *Annals of Oxford*, Oxford, N.Y., 1906, p. 424.
[2] Ibid.
[3] F. C. Soule, *The Chenango Canal*, op. cit., p. 7

its construction costs nor on its anticipated revenues alone, but "if the revenue promises to be sufficient to keep it in repair when finished, to defray the expenses of superintendence and the collection of tolls, and to meet the claims for interest on the capital expended, sound policy requires that it be constructed. Even if a less favorable result should be anticipated for a few years, the question of authorizing the construction of a public work may yet very properly be entertained. . . ."[4]

As if a flume gate had opened, approval for the canal swept the Legislature. The Canal Bill passed on February 23, 1833.

Almost immediately, as soon as riders could gallop from Albany with the news, victory celebrations dazzled the Chenango Valley. Bands played, fireworks glowed, cannon boomed, and citizens marched in a multitude of parades. Bonfires lit the streets of Norwich, and revelers there drank "dozens of champagne." Joyful residents dragged old river boats through the streets in Sherburne. At least 500 people attended a gala Canal Ball in Oxford.

John Tracy died in 1864, before the canal era ended. At his death Oxford had a population of over 1200. Workers at its canal docks and warehouses loaded the boats with crops from the nearby farms as well as products of the village's hoe factory, sash and blind factory, carriage shop, grist, saw and plaster mills.

Many years before, during the fight for the Chenango Canal, one legislator had prophesied naively that railroads "would never successfully compete with canals, but would become valuable tributaries to them."[5] In the 1870s the "valuable tributaries" completed their tracks through the Chenango Valley. "Honest John" did not live to see the demise of the canal and the birth of a new transportation system based on railroads. He never saw the days when weeds filled the towpath, and along the canal banks wildflowers clustered like bouquets massed against a monument to a mighty battle.

---

[4]Noble E. Whitford, *History of the Canal System of the State of New York,* op. cit., p. 679.
[5]James H. Smith, *History of Chenango County,* Syracuse, D. Mason & Co., 1880, p. 94.

Home of Lt. Gov. John Tracy, built about 1825.

St. Paul's Espiscopal Church, Washington Square, was built in 1857 on the site of the first Oxford home of one of the founders of the church, Benjamin Butler. The church was founded in 1815.

Construction of the present church began in 1856. Among the contributors to the building fund were Trinity Church, New York City, and Gov. Horatio Seymour, a former student at Oxford Academy. The church was finished in 1857 at a total cost of $13,387. The stone porch and bell tower were added in 1873.

The beautiful Waterford crystal chandeliers came originally from England before the Revolution and hung for many years in St. George's Chapel, New York City. In 1868 the daughters of Gerrit H. Van Wagenen, Oxford, arranged to have them installed at St. Paul's.

Among the church's lovely memorial gifts is the brass angel lectern presented in memory of Mrs. Julia Clapp Newberry, granddaughter of Benjamin Butler.

In the mid-19th century the house on Washington Square was built for the distinguished lawyer Henry R. Mygatt. After attending Oxford Academy, the young Mygatt went to Hamilton College for two years and graduated from Union College in 1830. He studied law in James Clapp's office, Oxford, and practiced law in this village throughout his lifetime. He married John Tracy's daughter, Esther.

For many years Henry R. Mygatt served on the board of trustees of Oxford Academy, the noted institution founded in 1791, chartered in 1794. At the time of the Academy's Jubilee, 1854, he was chairman of the board.

The Gothic Revival house on Washington Park was built in 1850.

The William Mygatt homestead was built in 1836. The construction date was authenticated during a late 19th-century remodeling, when the following information was found on an old window casing: ". . . Oct. 5, 1836, This day the ground is covered with snow and continues snowing 9 o'clock a.m. Six o'clock p.m. snowing yet. . . . Lemuel Lewis, Master builder of this house."

William Mygatt owned a large tannery. At one time he sold leather goods in a general store located in the homestead.

The residence at the southeast corner of Washington Sq. was known for many years as the Hyde house. The Hyde family owned the property as early as 1855.

Gerrit Huybert Van Wagenen built the house at the head of Washington Square in the early 1820s. A native New Yorker, son of a Dutch hardware merchant, he had served as lieutenant in the Revolution and had participated in the storming of Quebec. In 1821 he moved to Oxford with his wife, the former Sarah Brinckerhoff, and their large family.

Mr. Van Wagenen opened a hardware store in Oxford, carried on a business in buying and selling land, and built a grist mill, carding and woolen mill, and a saw mill.

The west wing of the house was removed by a new owner in the 1880s.

Squire Morehouse is known to have owned the house at 18 Albany St. in 1867. The site had been the property of the Morehouse family for many years prior to this date. It is probable that the Morehouses purchased land on Albany St. from the pioneer settler Uri Tracy.

The house on Fort Hill Park was erected about 1810 for Theodore Burr, probably the greatest designer of wooden bridges in the United States. The Burr truss which he invented and patented consisted of a system of panels with braced and cross-braced timbers, supported by large arches. His bridge over the Hudson River at Waterford, built in 1804, was the most notable bridge of its day in the United States. The bridge, built of yellow pine, was 797 feet long, had four spans, two roadways each of which was 11 feet wide. The bridge remained in use until 1909.

Burr built most of the early arch bridges across the Susquehanna, including the one at Harrisburg dated 1813. The Burr truss was used in the bridges which crossed the Chenango Canal.

Dr. Charles H. Eccleston, dentist, built the house at the corner of Ross St. and South Washington Ave. about 1853.

A tavern which stood on the site burned in 1851 while Dr. and Mrs. Eccleston were staying there on their honeymoon. The bride lost her entire trousseau. The gallant dentist is reported to have comforted his new wife with the words, "I'll build you a house on this site." The bride chose the design from an architectural plan in Godey's Lady's Book.

Oxford to Greene

The residence on Rte. 12 north of Greene was built with stone left over from the Chenango Canal locks in 1838, by D. S. Crandall. For many years Mr. Crandall ran a canal tavern, and prospered from traffic on the highway as well.

According to tradition the tavern was a stop on the Underground Railroad and gave shelter to escaping slaves. Furthermore, it provided a hiding place for stolen horses. The Loomis Gang, notorious Madison County horse thieves of the mid-19th century, are said to have hidden horses in the cellar, leading them down there on a ramp.

The veranda was added in the early 20th century.

The farmhouse on Elmer Smith Rd., off Rte. 12, north of Greene, was built about 1875 for George Chamberland, a buggy painter and carpenter. Mr. Chamberland married a Tillotson girl, member of a prominent family who lived nearby. According to one legend, the carpenter "went broke" trying to keep up with his more prosperous in-laws.

The lovely Italian villa on Elmer Smith Rd. north of Greene was built for S. Tillotson in 1855.

The Juliand homestead, 2 Juliand St., was built by Capt. Joseph Juliand in 1810.

HON. FREDERICK JULIAND

Frederick Juliand was five years old when, in 1810, the family moved the small distance from their cabin in the valley to their new home on the hill where the old French Road met the new branch of the Catskill-Susquehanna Turnpike. From miles away, turnpike travelers could see the big white house. Westbound settlers halted their oxen and spent the night before they plodded toward the rich lands of the Genesee country. Eastbound drovers rested as they trudged to Catskill, turning their herds into one of the three yards Frederick's father built across the road from the hospitable house. Often, Indians came in the night, and slept on the floor.

In 1798 the former French sea captain, Joseph Juliand, had brought his family from Connecticut to Greene, expecting to join the French refugee colony beside the Chenango River. He was disappointed. When the Julilands arrived, most of the French were preparing to move southward to Pennsylvania, disheartened and defeated by the wilderness. Although his countrymen left, Captain Juliand stayed in Greene, and he and his family helped make the settlement into a prospering community.

Frederick, the youngest of the Juliands' six children, received a classical education. He attended the Utica Academy, the Oxford Academy, and took extra Latin studies under a tutor in Greene. At 19 he moved to nearby Bainbridge for a few years and worked in a mercantile store. In 1830 he returned to Greene and went into partnership with his brothers George and Lewis. The Juliand Brothers partnership operated farms on the outskirts of the village and ran a mercantile store in the center of town.

When the Chenango Canal opened in 1837, farmers and storekeepers hurried to profit from the new opportunities. Farmers prepared to raise and market vastly larger crops. Storekeepers readied their shelves for a surge of new customers. The brothers dissolved their partnership. Lewis and George took over the farms, while Frederick became sole owner of the Juliand store.

Frederick was in his thirties. He was a slight, pleasant-looking man with a big smile. He wore his hair brushed forward over his broad forehead. His eyes were set wide apart. He dressed with elegance. In

1841 he married for the second time. The bride, Catherine, wore a wedding dress of orchid silk. Frederick's wedding coat was dark, with wide lapels, and he wore a white georgette cravat. When the couple left after the wedding, the bridegroom wore a top hat which carried a London label.

In 1856 Frederick Juliand went to Albany for his first term in the New York State Assembly. He was a Whig, believed in temperance, abhorred slavery. The official record of his term states that he served with distinction on the Committee on Banks. The unofficial record of that year whispers that he served with compassion to help fugitive slaves flee to Canada via the Underground Railroad.

By the year 1861, North and South were locked in the Civil War, and thousands of men fell in the terrible battles. At that time there was no draft. States recruited volunteer troops. In 1862 when President Lincoln made his anguished call for "300,000 more," Frederick Juliand was appointed to the recruiting committee for the 23rd Senatorial District, Chenango, Madison and Cortland Counties. At this time he gave the mercantile store to his son, perhaps in order to devote himself entirely to patriotic duty. He threw himself wholeheartedly into war work. With his efficient help the district recruited 1,000 men. The ten companies formed the 114th Regiment.

On September 6, 1862, the regiment assembled in Norwich, to leave for war. The recruits marched down Main Street. Many of the companies carried flags presented to them by their home villages. At sundown they embarked on ten chartered canal boats at the Curtis warehouse dock on South Main Street. The boats glided south to Binghamton, and in the villages along the way, little children leaned over the canal bridges to throw flowers on the soldiers.

In 1864 Frederick Juliand was elected to the New York State Senate. When he took his seat in the following year, a new, bright era seemed to be dawning for the Chenango Canal. The Chenango Canal Extension had been authorized two years before, after a very long struggle. It would build the Chenango 40 miles from Binghamton, via Owego, to Tioga Point near the Pennsylvania state line, where the New York State canal

would intersect the Pennsylvania canal system. The waterway had been a long-standing, magnificent dream. In 1838 one advocate had pointed out exuberantly that it would be "a connection between the two greatest and most extended chains of internal improvements in the world . . . would unite the Hudson, St. Lawrence and upper lakes with the Susquehanna and Ohio rivers and Delaware and Chesapeake bays."[1]

For many years after the opening of the canal the dream languished, but the demand for coal in the 1860s brought it back to life. The Extension could bring coal from Pennsylvania northward, without trans-shipment at Binghamton. Coal could float to the Erie and its laterals "to warm our hearths, supply our furnaces and forges, and propel our steamboats . . . and machinery of every description."[2] One New York State canal commissioner estimated that the canal with the extension would carry 200,000 tons of coal per year.

Senator Juliand hastened construction of the Extension. Following his decisive speech in the Senate, contracts for the first 10 months were let. Construction started in 1865. During the remainder of his term, he gave the Extension his "immediate supervision" and "active exertions."[3]

Prices for labor and materials soared in the post-war inflation. Construction costs multiplied. One more appropriation passed the Legislature and more contracts were let, but in 1870 twin lines of steel reached through the Chenango Valley and stopped the work. The railroads could tap the coal-fields directly, and the dream expired. The canal was abandoned, unfinished, in 1872.

Senator Juliand retired from public life after a final term in the Assembly. He lived to the age of 95, respected, admired and liked by his fellow citizens. In his "long, happy and successful life, he . . . won a reputation for purity of purpose, dignity of character, ability and enterprise," and he was remembered with tenderness and affection."[4]

[1]Whitford, op. cit., p. 696.
[2]Henry Wayland Hill, *Waterways and Canal Construction in New York State,* Buffalo Historical Society, 1908, p. 176.
[3]James H. Smith, *History of Chenango County,* Syracuse, 1880.
[4]Ibid.

"Princess" Go-Wan-Go's childhood home at South Chenango and Page Sts. was built in 1859. The noted actress and horsewoman headed a troupe of cowboys and Indians which performed to enthusiastic audiences in the U.S. and abroad. Go-Wan-Go, who added the title "Princess" to her stage name, was the daughter of Dr. and Mrs. Allen Mohawk and lived in Greene until she was nine years old. In 1870, after her father's death, she and her mother moved to Ohio.

The Gothic Revival house at 17 North Chenango St. was built about 1850, probably for Willis Sherwood.

Judge Robert Monell was the original owner of the house at 27 North Chenango St., built in 1838. Hon. Monell, who had settled in Greene in 1811, was a very public spirited citizen. He served five terms in the New York State Assembly, and three terms in the United States House of Representatives. He was Circuit Judge of the 6th Circuit, District Attorney of Chenango County, and president of the village of Greene.

The house at 11 Jackson St. was built by the carpenter Henry Beals about 1837.

Although many members of the Juliand family, prominent early settlers of Greene, lived in the village in the 18th, 19th and 20th centuries, the only residents today who carry the famous name are Mr. and Mrs. Frederick Juliand, the present owners of the house.

In 1829 the house on the south side of West Genesee St. was built for Dr. Augustus Willard. Born in Stafford, Ct., Dr. Willard settled in Greene in 1824 and practiced medicine in this village throughout his lifetime. He was president of the New York State Medical Society in 1857.

The Marcy homestead stands on the west side of Rte. 12, south of Greene. The Federal and Greek Revival style dwelling was built in 1829. It was acquired by the Marcy family in 1865, and is owned by that family still.

Chenango Forks

The house on the north side of Route 72 was built about 1817. A Chenango Canal lock is located below the house at the west. Some of the freight boat companies are said to have stabled their mules at a site near this lock.

The house at 807 River Rd. was built for the Middaugh or Tillotson family in 1862, on land owned originally by Silas Pepoon.

Binghamton

Christ Church, 187 Washington St., was built in 1853. It was designed by the famous architect Richard Upjohn of New York City, known in the mid-19th century as the foremost architect of Gothic Revival style in the United States.

The church stands on land given to the Episcopal Society by Gen. Joshua Whitney in 1806. The spire was added in 1906.

The original portion of Charles Samuel Hall's house, 171 Front St., was built in 1854. The house was enlarged in 1857.

CHARLES SAMUEL HALL

The opening of the Chenango Canal, May 6, 1837, drove Binghamton into passionate rejoicing. Bands played, cannon boomed, ladies fluttered their handkerchiefs, crowds trod the canal banks in a rush to welcome the "Dove of Solsville" and its distinguished passengers as the boat completed the inaugural trip down the canal from Utica.

Engineers and officials debarked amidst cheers. Orators predicted a shining future for Binghamton, starting immediately. No longer would local products be kept from New York State markets by the long, tortuous roads. The new waterway joined Binghamton to the prosperous activity of the Erie Canal. Boats loaded with lumber and farm products could start right away to move smoothly to Utica. From there they could glide along the Erie to Albany and join the big fleets of canal boats which the tow steamers pulled down the Hudson to New York City. The dawn of a new era so delighted the residents of Binghamton that the celebration continued for several days.

Even before the opening day, news of the completion of the canal had begun to pull settlers to Binghamton. Samuel H. P. Hall, a Connecticut merchant, had hurried his family westward to take advantage of the rich opportunities at the new canal terminus. The Halls had sailed along Long Island Sound and up the Hudson. From Albany they sped westward by railroad. At Utica they climbed into a four-horse stagecoach and jolted down to Binghamton on the rough roads. They reached Binghamton four days after the "Dove." Charles Hall, the oldest son, had just passed his tenth birthday.

In the years after the opening of the canal, Binghamton, like Charles Hall, grew fast. The village expanded as more settlers built houses on the new streets. It grew stronger as commerce, the muscle of a growing community, increased. In winter when the canal closed, the village boys used to skate. At first they glided along an avenue of ice which glistened between still, white fields. Each year more buildings lined the banks until the boys sped between rows of canal-side warehouses, shops and factories.

When Charles was 17 he went to Yale. It was a proud day for his father. Although one of his ancestors had been president of Harvard, one of Charles' grandfathers and two of his great grandfathers had graduated from Yale. In college the Binghamton student made a good record. He rowed in the boat races, studied hard, won several prizes. After graduation he entered Yale Law School and received his LLB degree in 1850. He lingered at New Haven for a year and got an M.A. degree in 1851.

As if steering his boat into the shining waters of a smooth, wide river, Charles Hall moved easily into a law career in Binghamton. He entered the office headed by the distinguished U.S. Senator Daniel S. Dickinson, and he plunged into village affairs with enthusiasm. He served on the Village Council, and as Village Attorney he helped draft the city charter which was adopted at the close of the Civil War. He was appointed Commissioner of the U.S. Circuit Court for the Northern District of New York.

Charles Hall's law practice flourished in a prosperous community. By 1851 the Erie Railroad had linked Binghamton to the Hudson River and had started to push farther west. At the same time, the Chenango Canal, like a trusted old friend, continued its familiar support. Timber and manufactured goods as well as farm products by the ton and bushel moved to market by canal from Binghamton warehouses. D. L. & W. Railroad cars brought coal from Pennsylvania to Binghamton where it was loaded on canal boats and hauled north. Thousands of tons of heavy bars of pig iron came south each year from Franklin, near Clinton, to be unloaded at Binghamton wharves.

Three years after he joined Senator Dickinson's office, Charles Hall became engaged to Mary Harris of fashionable Ballston Spa. She was a brilliant girl who had graduated with highest honors from Miss Willard's School in Troy. The young lawyer bought a handsome property on the west bank of the Chenango River, and he built the lovely brick house which stands today at 171 Front Street. He brought his bride to this house in 1855.

The Halls brought up their three sons in the house on Front Street. Proudly, they saw their eldest, Charles S., leave from here to go to Yale. In 1881 the Halls celebrated their silver wedding anniversary in the house, and in the following year Mary died.

Charles remained a widower for four years, and then he married the charming Annie Hastings Knowlton of Ohio. With his new wife he continued to live in the brick house with its big library, its broad lawns and its views of the Chenango River.

Proud of his illustrious ancestors, Charles Hall found time in his later years to trace the family tree from its deepest roots. He read and corresponded and researched with the avidity of a young lawyer, and he wrote two books. One dealt with the career of Samuel Holden Parsons, an ancestor who served as a general in the Revolutionary War. His major work, *Hall Ancestry*, followed the family history for over a thousand years and showed the Halls' relationship to emperors and kings, nobles and commoners. Charlemagne was a lineal ancestor on Charles' father's side, and his mother's family traced their ancestry to Hugh Capet, the 10th-century king of France. "Of the 25 barons chosen by the English nobles to enforce observance of Magna Carta," he wrote, "eight were lineal ancestors and the ninth was the son of a lineal ancestor"[1] He did not neglect the commoners. The book gave "some account of nearly one hundred of the early Puritan families of New England."[2] His own ancestors had taken part in the Puritan emigration of 1630–40. "That our early ancestors had a part, and some of them a very conspicuous part in this grand movement is something their descendants cannot fail to regard with increasing pride. . . ."[3]

Charles Hall died at home in 1910. He was, in the words of a Yale publication, "a felicitous writer, a delightful companion and an honored and useful citizen."[4] He earned and enjoyed a dignified prosperity. Probably, the ancestor who emigrated with his family to Binghamton at the opening of the canal would have been pleased.

---

[1]Hall Ancestry.
[2]Ibid
[3]Ibid
[4]*Yale Obituary News*, June, 1910.

Ezekiel Crocker was the original owner of the house at 1140 Chenango St., Hillcrest, built between 1790 and 1805. A shoemaker by trade, Mr. Crocker purchased land in the Clinton & Melcher Patent in 1790. He held New York State patents for 17 islands in the Chenango and Tioghnioga Rivers. The house faced one of his islands in the Chenango River. Mr. Crocker owned extensive tracts of land in addition to the islands, and was one of 60 proprietors of the 230,400-acre Boston Ten Towns tract. He became one of the richest men in Broome County, but lost his properties in unfortunate speculations and died poor in 1824. The house became the property of Ezekiel Crocker's son David.

Chenango Canal boatmen usually stopped at the "Big Spring" which bubbled on the Crocker property, and they filled their water barrels here. The water was said to keep better than any other along the canal route.

The house at 1314 Chenango St., Hillcrest, may have been built for Petrus Bevier, but more probably was built for William VanName and his wife, a daughter of Joshua Mersereau. The house dates from about 1800. Joshua Mersereau, with Petrus Bevier, had purchased a 358-acre tract from the Clinton & Melcher Patent in 1796. The house remained in the VanName family until about 1919.

The brick, Greek Revival house at 825 Front St. was built about 1850 for John Ward Cutler. In 1866 Mr. Cutler founded the Lake Ice Co. and cut ice from a pond behind his house and probably from other ponds as well. The business was large. Dormitories were built to lodge the Cutler employees. Mr. Cutler was a Supervisor of Broome County, 1867.

The Classic Revival house at 63 Front St. was built about 1840 for Franklin Whitney, one of twin sons, Washington and Franklin, born 1803 to Gen. and Mrs. Joshua Whitney.

As land agent for the Philadelphia merchant William Bingham, Gen. Joshua Whitney promoted the development of the tract of 30,600 acres which Bingham and his partners bought from New York State in 1786. From 1800 Whitney supervised the clearing of land for cabins at the junction of the Susquehanna and Chenango Rivers, and he laid out streets for a village. The settlement, called Chenango Point at first, became Binghamton.

John Stewart Wells, who built the house at 71 Main St. about 1870, was a prominent building contractor. His father having died when he was only six years old, Wells grew up on his uncle's farm near Marathon. When he was 17, he walked to Binghamton and learned the carpenter's trade. Soon he became one of the most esteemed contractors in the community. Among the notable buildings credited to him were Christ Church, St. Patrick's Church, the Congregational Church and the Erie Railroad Depot.

John Hilton Jones, a composer and professor of music, was the original owner of the Greek Revival house which stands today at 8 Riverside Dr. It was built in 1840 at 17 Front St. and moved to its present site about 1906.

Today's West Presbyterian Church Parish House, 85 Walnut St., was built about 1870 for Mr. and Mrs. William W. Hemingway. A versatile merchant, Mr. Hemingway started in the grocery business about 1854 soon after he first arrived in Binghamton. Subsequently he sold boots and shoes, hats and furs, lived in the West for five years, became a jobber of woolen goods in New York City, and finally returned to Binghamton where he entered the hardware business and became one of the most prominent and prosperous citizens of the community.

The weathervane adorns the carriage house at the former Hemingway residence, now the property of the West Presbyterian Church. The carriage house is located immediately behind the parish house.

Acknowledgments

The writer and the photographer wish to express their gratitude to those who helped them prepare this book. Dr. David Ellis, P. V. Rogers Professor of American History Emeritus, Hamilton College, honored us by his encouragement and paid us the high compliment of writing the Foreword. Clifford Lewis, Media, Pennsylvania, whose ancestors came from Oxford, suggested the idea of a book about the Chenango Canal route. Earl Widtman of Utica gave invaluable guidance.

Our sincere thanks go to all who helped us gather material about the canal, the homes, and the early homeowners. Arranged by counties along the canal route these include:

Oneida County, Ruth Auert, Eloise Beerhalter, Edward Bellinger, *Clinton Courier* files, Archibald Putnam Davies, Anne Douglass, Kenneth M. Fuller, Camilla Garvey, Frederick Griffin, Agnes Harding, Stuart Kellogg, Virginia Kelly, Frank K. Lorenz, Hamilton College Library, Kenneth McConnell, Richard N. Miller, Philip Munson, Munson Williams Proctor Library staff, Douglas Preston and the Oneida Historical Society, Silvia Saunders, Elizabeth M. Schmitt, Utica Public Library staff.

Madison County, Isabel Bracy, Harold Fleming, Madison County Historical Society staff, Susan Schapiro, Howard D. Williams.

Chenago County, Earlville Public Library staff, Mildred C. Folsom, Guernsey Memorial Library staff and librarians in Otis C. Thompson Local History Room, Mr. and Mrs. Maurice Ireland, Mr. and Mrs. Frederick Juliand, Thomas Lloyd, staff of Moore Memorial Library, Greene, Fay Pike, Mae Smith and the Chenango County Historical Society, Charlotte Stafford, Sumner Wickwire.

Broome County, Margaret Axtell, Larry Bothwell, Marjory Hinman, Roberson Museum staff.

We are indebted to the staff of the Canal Society of New York State and to Rose Baldwin, Wayland, Mass.

Bibliographical Essay

Canals

Authorities on the canals of the 19th century include Noble E. Whitford, *History of the Canal System of the State of New York,* 1906, and Henry Wayland Hill, *Waterways and Canal Construction in New York State,* Buffalo Historical Society, 1908. Also, Alvin F. Harlow, *Old Towpaths,* 1926, Harry Sinclair Drago, *Canal Days in America,* 1972.

Material on the Chenango Canal is very limited. For its general history see Barry Beyer's comprehensive *Chenango Canal,* 1954. Construction and operation of the canal are described in F. C. Soule, *The Chenango Canal,* compiled for Canal Society of New York State, 1970. See also, Captain Frank H. Godfrey's *Letters* published by the Canal Society 1973, and the Chenango County Planning Board's *Chenango Canal.* The Canal Society, Syracuse, has some newspaper accounts in its files. The Guernsey Library, Norwich, has a rich store of early newspapers on microfilm, and many of these contain accounts of the battle to win authorization for the canal.

General Histories of the Area Which Include Material about the Original Homeowners

Oneida County: The early authority is Pomroy Jones' *Annals & Recollections of Oneida County,* 1851. Some biographical material is available in Evarts & Fariss, pub., *History of Oneida County,* 1851, as well as Daniel E. Wager, *Oneida County,* 1896, the county's new *Oneida County,* 1977, and Durant's *Oneida County,* 1878.

Utica has M. M. Bagg's two splendid volumes, *The Pioneers of Utica,* 1877 and *Memorial History of Utica,* 1892. Dr. T. Wood Clarke's *Utica For a Century and a Half,* 1952, is a valuable reference. Judge John J. Walsh's *Frontier Outpost to Modern City,* 1978 and *Vignettes of Old Utica,* 1975, are filled with fascinating material. Blandina Dudley Miller's *A Sketch of Old Utica,* 1913, gives an account of some early residents. The files of the Oneida Historical Society contain a quantity of Alfred Munson's correspondence, as well as material about the committee which investigated the early steam mills.

Clinton has an early authority in the Rev. A. D. Gridley who wrote the *History of the Town of Kirkland,* 1874. Helen Neilson Rudd's *A Century of Schools in Clinton,* 1964, has detailed accounts of the early schools. Other excellent books about Clinton residents are Walter Pilkington, ed., *Journal of Samuel Kirkland,* 1980, his *Hamilton College,* 1962, and his little booklets *The Kirkland Cottage* and *The Homestead.* Grace C. Root's *Fathers and Sons,* 1971, is a sensitive appreciation of the Root family. Edward W. Root's *Philip Hooker,* 1929, is an excellent appreciation of the great architect. Philip C. Jessup's *Elihu Root,* 1938, is a comprehensive and authoritative account of the statesman's life. Mary Bell Dever's *History of Clinton Square,* 1961, has some material, as does *Clinton, N.Y. and Vicinity,* compiled by a class in American History, Clinton High School, 1912, and the *Clinton American Revolution Bicentennial Booklet,* 1976.

Deansboro is fortunate in having Thomas Dean's journal, *A Voyage to Indiana in 1817,* reprinted by Town of Marshall Bi-Centennial Committee 1976, and Kenneth G. McConnell's *History of the Town of Marshall,* 1976.

David Beetle's *Along the Oriskany Creek*, 1947, is colorful and fascinating. For the Indians in the South and the Trail of Tears, see S. E. Morrison, *Oxford History of the American People*, 1965, vol. 2, Chapter VIII.

Oriskany Falls has *The Colonel's Hat*, a history of the township of Augusta, published 1976.

Madison County: The standard works are John E. Smith, *Madison County*, 1899, and *History of Chenango and Madison Counties, 1880*, as well as Mrs. L. Hammond-Whitney's delightfully personal *History of Madison County*, 1872. Isabel Bracy has compiled *Odds and Ends of History of Madison County*, which includes engaging vignettes of early settlers. Madison County Planning Board's *Country Roads*, 1976, is a very valuable reference, with photographs of early buildings and brief historical sketches of the villages.

Eaton has Louise Isbell's *Eatonbrook Valley*, 1970, and the Town of Eaton Bicentennial Committee's *The Town of Eaton in the Bicentennial Year*, illustrated with early photographs, and William R. Houghton's *Selected Topics of Rural Historical Interest and Explanatory Drawings*, 1971. Harry E. Hart's *Last Gathering*, 1972 and his "Lebanon Hill Journal" vol. II, 1976, contain some fascinating material about the early families.

Hamilton is rich in material, with Howard D. Williams' authoritative *A History of Colgate University*, 1969. The pamphlet *Historic Hamilton, a Walking Tour*, 1979 and *Hamilton Walk Book*, 1973 contain detailed local information. *Progressive Hamilton*, 1896, is a good history of the village.

Earlville has a *Condensed History* compiled from articles by J. R. Parsons, 1938, as well as a speech by Bill Lawrence Rabson in 1951, *A History of Earlville*. Donna Burns, *The Letters of Charlotte Browning Page*, 1980, contains some material about the Page family of Earlville. The Earlville Library has scrapbooks of early newspaper articles about the area.

Chenango County: See Chenango County Historic Building-Structure Inventory. Standard histories are Smith, *History of Chenango and Madison Counties* and his *History of Chenango County*, 1880, and Hiram Clark, *History of Chenango County*, 1850.

Norwich is well documented in Albert and Goldie Phillips' two slim volumes, *Annals of Norwich*, published by the Norwich Historical Society in 1964 and 1965, and Louise S. Shinners' *Norwich Golden Anniversary*, 1964. The files of the Guernsey Memorial Library contain biographical material about some of the early residents.

Sherburne is covered by colorful anecdotes in George Walters, *Sinners & Saints*, 1973, *Chenango Valley Tales*, 1962, *Chips & Shavings*, 1966, as well as Roy Gallinger, *Campfires in the Forest*, 1966.

Oxford's standard work is Henry J. Galpin, *Annals of Oxford*, 1906. Some biographical material is found in Rev. Roland A. Boutwell, *Some Early Settlers of Oxford Before 1800* published in 1966, and the *Oxford Academy Jubilee*, 1854.

Greene has Mildred English Cochrane, *From Raft to Railroad*, 1967, *Milestones, 1776–1976*, and Mildred English Cochrane Folsom, *Annals of the Town of Greene*, 1971, all very helpful and interesting.

Broome County: The standard histories are William S. Lawyer, ed., *Binghamton*, 1900, J. B. Wilkinson, *The Annals of Binghamton of 1840*, re-published 1967, *Biographical Review of Leading Citizens of Broome County*, 1894, Mather & Brockett, *A Geographical History of the State of New York*, 1852, and Dr. George Burr, *Historical Address*, 1876.

Binghamton's buildings are described in *Historic Architecture of Broome County and Vicinity* by Eugene D. Montillon, 1972, and *Historic Preservation in Broome County* published by Broome County Department of Planning, 1979. The Roberson Museum, Binghamton has the *Sun Bulletin*'s series of articles on the history and growth of Binghamton, 1976, and it has a copy of *Hall Ancestry* by Charles S. Hall, 1896.

Architectural Styles

See *American Buildings and Their Architects*, by William H. Pierson, 1980, *Wood & Stone, Landmarks of the Upper Mohawk Region*, 1972, *A Field Guide to American Architecture* by Carole Rifkind, 1980, *Country Houses* by A. J. Downing, 1950 (Gothic Revival), *A Home For All*, by Orson S. Fowler, 1954 (Octagons).

New York State History

Anyone who wants to try to understand the canal era should consult *A History of New York State* by Ellis, Frost, Syrett and Carman, Revised ed., 1967, Chapter 20.

Designed by Frank Devecis, typography in Palatino typeface and printing by Canterbury Press of Rome, New York. The superb photographs were reproduced in a duotone process and printed on 80 # Centura Dull paper for the text pages. The binding was by Riverside Bindery of Rochester.